D0448838

DATE DUE

AP 9 '93			
JE 23 '94			
MR 17 '95			
MR 1 '96			
MY 3 '96			
NO 19 '96			
AP 17 02			
SE 2 5 09			
DE 4 12			

DEMCO 38-296

THE SKYWATCHER'S HANDBOOK

NIGHT AND DAY
WHAT TO LOOK FOR IN THE HEAVENS ABOVE

CONSULTANT EDITOR
COLIN A. RONAN MSc FRAS

CROWN PUBLISHERS, INC.
NEW YORK

Colin A. Ronan, MSc FRAS is the
editor of the Journal of the British
Astronomical Association and President
Elect of the Association, and was
recently a member of the Council of The
Royal Astronomical Society. He is a
trained astronomer and historian of
science and author of more than 25 books
– one of which (a biography of Edmond
Halley) has become a standard work. At
present he is working on a multi-volume
abridgement of Joseph Needham's *Science
and Civilisation in China*, being
published by Cambridge University Press.

Storm Dunlop, FRAS FRMetS is also a
member of the American Association of
Variable Star Observers. He is a writer on
astronomy and meteorology and has also
completed a major translation from the
German, *Variable Stars*. His other
interests are planetary geology and
meteorological photography.

Brian Jones is an amateur astronomer
who has edited a variety of astronomical
journals, including the *Handbook for
Astronomical Societies* for the Federation
of Astronomical Societies, and
contributes articles on astronomy to
popular-interest magazines. His recent
work is aimed at fostering an interest in
astronomy among a younger audience.

Editor: Jonathan Hilton
Managing Editor: Ruth Binney
Assistant Editors: Louise Tucker
                   Gwen Rigby
Art Director: Paul Wilkinson
Picture Editor: Zilda Tandy
Production: Barry Baker, Janice Storr

Compilations, design and text
The Night Sky © 1985, 1989, Marshall Editions Ltd
The Daylight Sky © 1985 Storm Dunlop
Observing and Recording the Sky
© 1985 Brian Jones
Picture credits © 1985 as acknowledged

Published in the United States of America by
Crown Publishers, Inc, 225 Park Avenue South,
New York, New York 10003

Conceived, edited and designed
by Marshall Editions Limited
170 Piccadilly, London W1V 1DD

Library of Congress Cataloging in
Publication Data
Main entry under title:

The Skywatcher's handbook

    Includes index.
    1. Astronomy—Observers' manuals. I. Ronan,
Colin A. II. Dunlop, Storm.
QB64.S58 1985   520   85-3810

ISBN 0-517-573261

10 9 8 7 6 5 4 3 2 1

First paperback edition

Printed in Belgium

C O N T E N T S

Introduction   6

**Part I THE DAYLIGHT SKY
STORM DUNLOP**

The sky above   10
The sky's colours   14
Optical phenomena   20
Wind movements   30
Water in the atmosphere   40
Cloud watching   44
Cloud varieties   46
Nature's fireworks   60
The Sun   70
Mirages   78
Nacreous and noctilucent
  clouds   80
Aurorae   82
The sky from above   84
Weather lore   90
Building a weather station   92

**Part II THE NIGHT SKY
COLIN A. RONAN**

Approaching the sky   100
The Solar System   102
Mercury and Venus   104
Mars   106
Jupiter   108
Saturn   110
Uranus, Neptune
  and Pluto   112

The Moon   114
The stars   120
The Galaxy   124
The constellations   132
Star magnitudes   138
Star watching   142
Guidance for observers   158
Nebulae and clusters   160
Double and variable stars   166
Shooting stars and comets   170
Artificial satellites   176

**Part III OBSERVING &
RECORDING THE SKY
BRIAN JONES**

The naked eye   182
Photographing the
  daylight sky   184
Photographing the
  night sky   188
Binoculars   192
Telescopes   196
Telescope mounts
  and drives   198
The home observatory   200
Keeping records   202
Appendices   204
Glossary   216
Index   220
Acknowledgements   224

The sky remains the most reliable source of the best things in life that are free. The Moon does, indeed, belong to everyone. So do rainbows, shooting stars, eerie eclipses and breathtaking sunsets. Although the sky's most dramatic phenomena do not occur every day, it is still true that, wherever you are, what takes place in any 24-hour period in the sky will provide substantial interest to the skywatcher who has learned to interpret it.

Observation of the sky was vital to ancient people – not only the learned, but shepherds, farmers, fishermen and sailors, who relied on the information it gave them day and night. Even today, these same groups of people have a vested interest in skywatching, and are enviable for the knowledge that seems just to come naturally. But we can all share in it if we want to. The sky is a recreational resource far too many people miss out on. Anyone who finds himself remarking on the weather (and that means everyone) is a potential skywatcher.

The two great sciences arising from the study of the skies are meteorology and astronomy. The original word for meteorology was coined by Aristotle to describe atmospheric conditions, and now relates particularly to weather forecasting. Astronomy derives from another ancient Greek word meaning star-arranging. It is generally associated with the sky at night – when heavenly bodies are most easily observed. Both studies can, of course, be a life's work; but equally, both can make rewarding leisure pursuits, open to all. A healthy curiosity and a little patience are enough to qualify you for skywatching.

Just as the Earth has its seasonal rhythms, so the sky has its own distinct pattern, with characteristics unique to dawn, full daylight, evening and dark. The first two sections of this handbook are devoted to exploring the day and night skies. The Daylight Sky will give you the basics of meteorology so that you can make sense of weather maps, identify the form and movement of the clouds and the direction of the wind, and attempt some predictions of your own. Here you will find everything from building your own weather station to the significance of a full solar eclipse.

In The Night Sky, even the city dweller, plagued by artificial lighting, will find help in picking out the more spectacular astronomical sights, from the readily perceived phases of the Moon to the changing pageant of the constellations. Familiarity with the planets is enormously satisfying; Mercury and, especially, Venus (the Morning and Evening Stars) can be magnificent in twilight and early evening. Jupiter, too, is a wonderful sight in the night sky, considerably enhanced by viewing it through binoculars. Meteors – the shooting stars of earlier times – comets, galaxies and the multitude of stars that can be seen with a telescope or the naked eye are all dealt with in The Night Sky.

The third section of this handbook concerns itself with Observing and Recording the Sky. Your skywatching takes on greater meaning when you make your observations permanent; they could be of real scientific value (comets West, Bennett and Alcock are among those bearing the names of their amateur discoverers). Here we describe the best ways to photograph the day and night sky, as well as how to build and equip your own home observatory.

The more you skywatch, the more useful you will find the Appendices, which are full of key dates for celestial events, essential statistics and even details of societies of amateur meteorologists and astronomers you would be welcome to join. Their enthusiasm and assistance can enrich your skywatching considerably. The sky is, in every sense, an immense subject, but it is also a generous and constantly accessible one. It is hoped that with the help of this book you can meet it half-way.

Colin A. Ronan, MSc FRAS

7

SKY

The sky above/1

It is hard to believe, when looking up into the sky, just how thin is the layer of atmosphere surrounding the Earth. The 'top' is usually described as being about 60 to 80 km (38 to 50 mls) above the surface, only about one-hundredth of the Earth's radius at the Equator – 6,378 km (3,963 mls).

There is, in fact, no true upper boundary to the atmosphere, but there is a transitional region where it merges into interplanetary space. This region is greatly affected by changes that occur during the solar cycle (see pp 70–7).

Atmospheric pressure is an important factor in governing the behaviour of the weather and also in its prediction (see pp 90–1). This pressure is the weight of the column of air above any particular point and is measured by one of the various types of barometer. Pressure was originally described in terms of the height of the column of mercury it would support, but it is now defined in scientific units as the force exerted on a given area. Atmospheric pressure is normally measured in millibars (mb), or thousandths of a bar. One bar is equivalent to the pressure exerted by 750.06 mm (29.53 in) of mercury at 0°C (32°F). Although pressure is by no means constant throughout the atmosphere, the average at sea level is about 1013 mb.

In the atmosphere as a whole, pressure declines upward, at first rapidly and later much more gradually. When you fly in an aircraft at a height of about 9 km (approximately 30,000 ft), the outside pressure is around 350 mb. This is a measure of the amount of air above you – roughly 65 per cent of the atmosphere lies below.

Temperature also varies throughout the atmosphere and the way in which it changes with height is used to define the various layers. The lowest layer, the troposphere, contains most of the mass of the atmosphere and here, too, the changes that form the

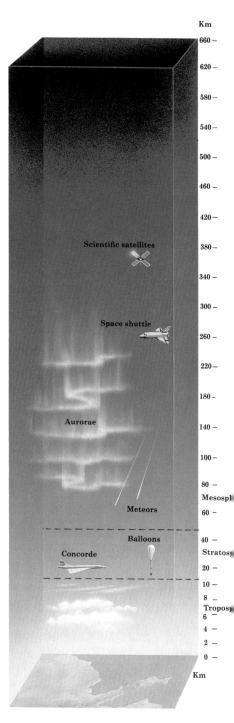

Km
660 –
620 –
580 –
540 –
500 –
460 –
420 –
380 – Scientific satellites
340 –
300 –
260 – Space shuttle
220 –
180 –
140 – Aurorae
100 –
80 –
Mesospl
60 – Meteors
40 – Balloons
Stratos
20 – Concorde
10 –
8 –
Tropos
6 –
4 –
2 –
0 –
Km

▷

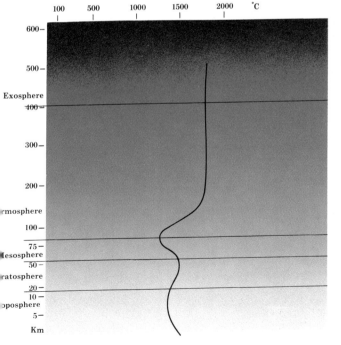

The boundaries between the various atmospheric layers, *left*, are marked by temperature reversals. At the highest levels, atoms in rarified (thin) gases move with great velocity, producing high temperatures.

100   500   1000   1500   2000   °C

600 —

500 —

Exosphere
400 —

300 —

200 —

rmosphere
100 —

75 —
Mesosphere
50 —
ratosphere
20 —
10 —
oposphere
5 —

Km

The depth of the Earth's atmosphere is shown, *above*, as a thin, blue band along the horizon.

Most clouds are confined to the troposphere at the bottom of the atmosphere, *opposite*. Here the dominant weather processes are also found.
   The boundaries between the various atmospheric layers, *left*, are marked by temperature reversals. At the highest levels, atoms in rarified (thin) gases move with great velocity, producing high temperatures.

11

The sky above/2

weather patterns of the world largely take place.

The tropopause, the boundary between the troposphere and the stratosphere, is usually fairly distinct, although its height may vary considerably. It is normally found at altitudes of about 18 to 30 km (11 to 19 mls) over equatorial regions and 8 km (5 mls) over the poles. Variations occur with the seasons – especially at higher latitudes – but the decline in height from Equator to poles is never uniform. There are always major 'steps' between the levels over the tropics and those over the middle latitudes, as well as lesser ones nearer the poles and many other irregularities. The locations of these breaks in the tropopause have a great effect upon the development of weather systems.

The Earth is unique among the planets in that its atmosphere largely consists of the heavy gases, nitrogen (78 per cent) and oxygen (21 per cent), with mere traces of other gases. The relative proportions are fairly constant up to about 80 km (50 mls), but the heavy gases become rare at greater heights and only hydrogen and helium are found in the outermost regions. In the stratosphere there is a high concentration of ozone at about 25 km (15 mls). This acts as a shield against damaging ultraviolet radiation from the Sun, only a small fraction of which actually reaches the Earth's surface.

Another important constituent of the atmosphere, water vapour, varies in concentration from place to place and from time to time. Because the tropopause is a strong inversion (a condition in which a layer of warm air holds cooler air near the Earth's surface); water vapour is largely trapped below it. For this reason, nearly all the clouds you can see are in the troposphere. Some cloud formations may stretch up to the tropopause, but in the stratosphere the air is dry and clouds are rare.

Setting an aneroid barometer
If you want to compare your barometer readings with those at other places, check the instrument's setting from time to time and adjust it if necessary. To do this, choose a day when the weather is fairly quiet (when the pressure is not changing rapidly), contact the nearest meteorological station and ask for the current pressure reading. Set the pointer on your barometer to this pressure, using the adjusting screw on the back of it. If the station is far away there may be a small pressure difference, but this will be insignificant. The pressure you are given will also be corrected to mean sea-level, so you will not have to worry about the drop in pressure with increasing altitude.

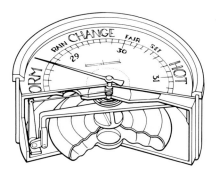

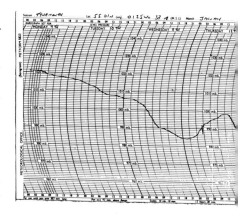

In an aneroid
barometer, *left*, a
closed, evacuated,
metal-walled
capsule collapses
with a rise in air
pressure and
expands with a
fall. This motion is
amplified to move
an indicator needle
over a circular
scale.

A barograph,
above, uses a stack
of aneroid capsules
to produce
movements of a
pen arm in
response to
changes in air
pressure.

In a mercury
barometer, *right*,
an inverted tube of
mercury is
arranged so that it
dips below the
surface of a
reservoir of
mercury. When the
level of mercury
drops, a vacuum is
created in the
closed, upper end.
Changes in air
pressure on the
surface of the
reservoir affect the
length of mercury
in the tube, which
is calibrated and
from which the air
pressure is read.

A trace from a
barograph, *left*,
shows the passage
of three
depressions.

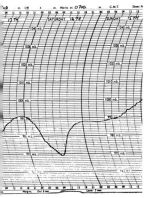

The sky's colours/1

Many of the colours seen in the sky, such as those in rainbows, are caused by refraction. When light moves in any transparent medium, such as air or water, a change in density (or transfer to another medium) causes a tiny change in the velocity of the light. Unless the surface is precisely at right angles to the beam of light, this is accompanied by a change in direction. The result is, for example, that a drinking straw placed in a glass of water appears to bend.

Light can be envisaged as a series of waves, the distance between consecutive troughs or peaks of which (the wavelength) determines colour. Blue light has a shorter wavelength than red. The extent to which refraction causes a change in direction depends on the wavelength of the light, blue light being deflected more than red. So the Sun's 'white' light, which contains a range of wavelengths, is dispersed into a spectrum by the process of refraction.

In ice crystals and raindrops in the atmosphere, and in a glass prism, the dispersion of colours may be readily visible. With window glass, however, the light leaves the higher-density medium through a surface parallel to the one by which it entered. Here the dispersion is cancelled out, so the final image is colourless. If the glass is tilted, the position of the image shifts slightly. Layers of air with different densities have the same effect and, as they fluctuate, the images of objects seen through them also appear to move. This motion can be seen in the shimmering of hot layers of air close to any heated surface.

The twinkling, or scintillation, of stars arises in the same way. With layers of air, however, the top and bottom surfaces may fluctuate independently in a series of waves so that the angles between them also vary, giving rise to colour changes. The result is most clearly seen at night,

Scattering is responsible for the range of tones, seen in the sky. The smallest water droplets in the cloud scatter light and appear white; larger droplets absorb more light and so appear darker.

Apparent position of the Sun

Actual position of the Sun

The atmosphere acts as a lens, changing the direction of light and making objects appear higher in the sky. On the horizon, this difference amounts to about half a degree – roughly the apparent diameter of both the Sun and the Moon.

▷

Spectral effects produce some of the sky's colours. Light disperses into a spectrum on passing through a prism. The gradation is smooth, but the eye sees a series of individual colours – red, orange, yellow, green, blue and violet. These pure, spectral colours are described as saturated.

The sky's colours/2

when bright stars close to the horizon often show brilliant flashes of colour.

The normal background colours of the sky – the varied blues and yellow and red tones at sunrise and sunset – arise from a completely different effect, known as scattering. This is caused by suspended particles in the air and even by the molecules of the air themselves. Without scattering, the sky would appear black, just as it does in space.

The amount of scattering depends strongly on the size of the particles and the wavelengths of the light concerned. Particles much smaller than the wavelength of light do not cause any scattering of the long (red) wavelengths, but toward the short (blue) end their scattering efficiency rises sharply. The size of gas molecules in the atmosphere is roughly one-thousandth of the wavelength of visible light, and they scatter light in all directions. It is this that gives rise to the blue of the daytime sky.

At sunrise and sunset, light from the Sun has to follow a long path through the dense lower atmosphere. Here, blue light is gradually removed by scattering, and the colour of the light shifts through yellow and orange until, finally, red wavelengths only remain. This is why the Sun and Moon appear red when close to the horizon.

The effects of scattering are often visible in thin smoke. Look away from the Sun, and the smoke appears to be blue; look in the opposite direction, toward the light, and the smoke appears brownish. With thick smoke, the situation is more complicated, since multiple scattering is involved. This gives rise to white light, whatever the size of the particles, so the centre of a column of thick smoke appears whitish, while the thin outer edges are tinged with blue.

Particles that are larger than the wavelength of light scatter all colours equally well, so the light appears

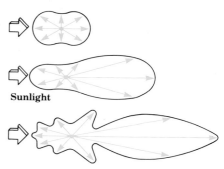

Sunlight

Forward scattering of light increases as the size of atmospheric particles increases. This produces a whitish patch centred on the Sun. This area of scattered light is missing only when the air is particularly clear and free from particles.

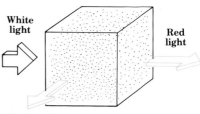

White light

Red light

Scattered blue light

The sky looks blue due to light-scattering effects. Although air is transparent, the vast number of molecules of gas in even a small volume of the atmosphere scatter blue light in all directions. These molecules have practically no effect on red light, which passes straight through. In most parts of the sky, away from the direct light of the Sun, only scattered blue light can reach an observer's eyes.

▷

Crepuscular rays can be seen when the air is full of particles and the light source is hidden. In reality, the beams of light are parallel, but they appear to diverge because of the effects of perspective.

Distant objects appear lighter and bluer than those near at hand. This is due to aerial perspective. It is caused by the combined effects of the scattering of blue and white light. Everybody instinctively uses aerial perspective to help gauge distance.

The sky's colours/3

white. This happens in clouds where the size of the water droplets and ice crystals is about 50 times that of the wavelength of light. Strongly illuminated clouds, therefore, appear brilliantly white. Their bases and other clouds in their shadow may appear dark grey, as would the same clouds if they were viewed against the light.

Clouds that are equally illuminated often appear as different shades of grey, and this gives an indication of the relative sizes of the cloud droplets. Larger droplets absorb more light, making the clouds look darker. Clouds may appear any colour, depending upon the colour of the light that illuminates them. Around sunset and sunrise, a whole succession of yellow, orange and red tints can be seen on the same cloud.

Sometimes a layer of clouds may cover the whole sky except for a narrow band close to the horizon. This strip of light is often strongly coloured, even when the Sun is high in the sky. Light passing under the clouds from the clear sky starts its journey as white but undergoes the usual blue-scattering on the way, becoming progressively redder. When there are gaps in the cloud layer at different distances from an observer, the light on the horizon may appear green, say, in one direction and orange in another.

Many types of particle contribute to atmospheric haze. The most common natural ones are dust, salt crystals, and pollen. Man-made pollutants may make a significant contribution at certain times and places. It is scattering by these particles that causes even a cloudless sky to become paler toward the horizon, where it may appear almost white. Sometimes the particles produce a whitish veil over the whole sky, which even overhead may be only an extremely pale blue. After rain has washed the atmosphere clean, the sky becomes a much deeper blue.

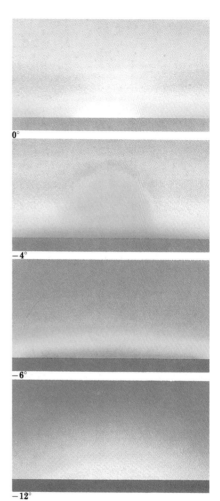

At sunset, *above*, a succession of colours can be seen in a clear sky. The numbers indicate the distance of the Sun below the horizon in degrees. Sometimes red light passing through the lower layers of the atmosphere may combine with scattered blue light to give rise to a vibrant purple light. While these changes take place in the west, a pink 'counter-twilight' may be noticeable in the east. This can appear as an Alpine glow in mountain regions.

Alpine glow

Purple light

Optical phenomena/1

A rainbow, probably the most common of all the optical phenomena of the daylight sky, is formed when sunlight illuminates falling rain. Light enters the drops and is dispersed into a spectrum, but at the same time it is internally reflected at least once before it leaves the drops. Although the light leaves every drop in all directions, there are strong concentrations at certain fixed angles, determined by the number of internal reflections.

The most common bow – the primary bow – is caused by a single internal reflection and shows the normal sequence of spectral colours. A secondary bow, produced by two internal reflections, often accompanies the primary; it is quite strong, lies outside the primary and has a reversed colour sequence. Both bows appear in the part of the sky opposite the Sun. A third-order bow is also possible, but it would lie on the other side of the observer, surrounding the Sun, and so it is unlikely it has ever been seen.

When you see an ordinary rainbow, notice that it appears to be part of a perfect circle and that its centre lies below the horizon by an amount that is always exactly the same as the Sun's altitude above it. Rainbows are constant in size, so the higher the Sun, the lower the top of the bow. If the Sun is too high, rainbows cannot be seen, and, conversely, a full semicircle may be visible at sunrise or sunset. If, as frequently happens, portions of the arc are missing, this means either that no rain is falling in that place or that the rain is not illuminated by the Sun, perhaps because it is in the shadow of another cloud.

Sometimes there are additional, supernumerary bows, lying inside the violet edge of the primary. These are normally brightest near the top of the arc, but overall their colours are more subdued than those of the two main bows, and they usually appear violet,

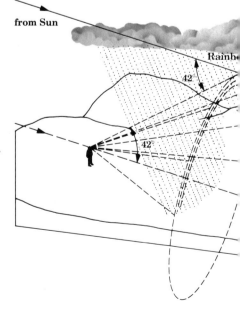

from Sun

Rainb

42

42

▷

Morning or late afternoon are the best times to see rainbows. In the photograph, *above*, the Sun is about 30° above the horizon. When the Sun is high in the sky, in the middle of the day, the tops of rainbows are low and may be difficult to see.

Rainbows always appear on the side of the observer opposite to the Sun. They are portions of circles centred on the antisolar point. The radius of the primary bow, *left*, is about 42°.

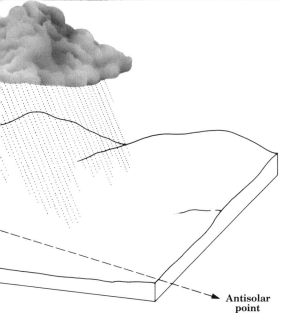

Antisolar
point

21

Optical phenomena/2

pink and greenish. Supernumaries are formed because two rays of light may follow slightly different paths within a raindrop but still emerge in the same direction. This gives rise to interference between the two rays and produces the alternating bright and dark bands in the supernumerary bows.

Far more striking are the reflected bows that occasionally occur if there is a sheet of water behind you when you observe a rainbow. You may then be lucky enough to see a second set of arcs, similar in size to the normal ones, but with their centres higher in the sky. Frequently, only the lower parts of these arcs are visible.

Rainbows do not always show a full range of colours. If the Sun is low, only orange or red light may remain, so these are the only colours you will see. Rainbows caused by moonlight – which are much fainter than ordinary rainbows – are often reported as being white. This, however, merely illustrates the limitation of our eyes, which are insensitive to colours at low light levels. Photographs show the usual set of colours.

The intensity of the colours gives an indication of the size of the raindrops. With large drops, more than 1 mm in diameter, violet and green are bright and red is distinct, but there is little blue. As droplet size decreases, so does the strength of the red. After passing through a stage when only the violet persists, the bow finally becomes white; in such a fogbow, the water droplets are below 0.05 mm in diameter. Supernumerary bows shift through yellowish shades and also eventually become colourless.

Rainbows are widely believed to give some indication of the forthcoming weather. They are not, however, at all reliable in this regard, and show only the conditions prevailing at the time – that rain is falling and the Sun shining. There is some truth in the saying that a rainbow seen in the

▷

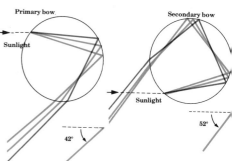

Primary bow

Sunlight

42°

Secondary bow

Sunlight

52°

A rainbow is created when one internal reflection, *above left*, in a raindrop returns and concentrates the light at an angle of about 42° to its original path. Two reflections form a secondary bow, *above right*.

A bright rainbow over the aerials of the Very Large Array (VLA) radio telescope in New Mexico.

In a primary bow, raindrops high in the sky send red light toward an observer. Violet light comes from lower drops. The sequence of colours and circumstances are reversed in a secondary bow. The space between the bows is known as Alexander's dark band.

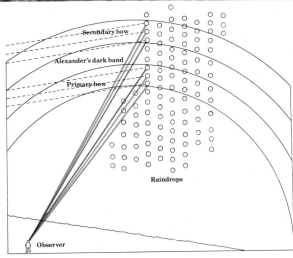

Secondary bow

Alexander's dark band

Primary bow

Raindrops

Observer

Optical phenomena/3

morning indicates that more rain is on the way, and that when one is seen in the evening, the rain will not persist. This is because a rainbow seen in the morning lies in the west, and in the middle latitudes many weather systems, including rain showers, approach from that direction. Similarly, a rainbow seen in the east probably shows that a particular rain shower has passed by. There could, however, be more rain approaching.

Another class of optical effect is caused by diffraction. You can see an example of this by making a small hole (no bigger than 0.5 mm in diameter) in a piece of cooking foil. When held up to a strong light, a set of narrow, closely spaced rings should be visible surrounding the central hole. The pattern of bright rings and dark spaces is caused by interference.

Similar phenomena, known as coronae, occur in clouds containing water droplets of uniform size (or even ice crystals); in this instance, yellow-white circles are produced around the Sun and Moon.

Do not confuse this form of corona with the Sun's outer atmosphere, which is known by the same name. The corona around the Moon is more commonly seen because the Sun is too bright to view directly without risking permanent eye damage.

When the Sun is covered by cloud, you may view its reflected image in a pool of water or in a piece of dark glass, or else wear dark glasses. As an added precaution, stand so that the Sun's disc is just hidden behind some suitable object, such as a telephone pole or a street sign.

Immediately surrounding the Sun (or Moon) is the aureole, a series of rings that are bluish in the centre and reddish-brown toward the outside edge. The diameter of the outer ring may vary considerably, but is frequently about 2°. Outside the aureole, one to three sets of coloured rings may

Halo phenomena take many forms and the most common only are shown, *above*. Some or all of these may be visible at any one time. The circumzenithal arc is more strongly coloured than the circular haloes.

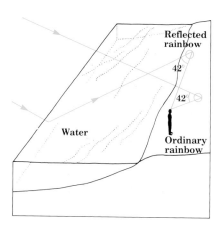

Sunlight reflected from water behind the observer, *above*, causes raindrops to be illuminated from a low angle. This gives rise to primary and secondary bows that show exactly the same colours and arrangement as the normal ones, but appear much higher in the sky. The more common 22° halo can be seen, *opposite*.

▷

Optical phenomena/4

be seen, always with the blue inner-most and the red outermost. The size of these coronal rings depends on the droplet sizes in the cloud – the purest colours are visible when all the drops are the same size. Variation in drop diameter causes the rings to overlap and become less distinct. The colours also become mixed, or less saturated, unlike those in a true spectrum.

A closely related phenomenon occurs in iridescent clouds. Such clouds – especially when they are thin – may be completely covered by beautiful bands of colour and may extend from close to the Sun to distances of 30° or even more. Sometimes the colours are extremely pale, or unsaturated, but on other occasions they may be brilliant, pure spectral colours. This effect, too, is often missed because of the brightness of the Sun.

In thicker clouds, the bands of iridescence often run approximately parallel to the edges. Here again, this gives a clue to the size of the water droplets, which tend to evaporate, and hence grow smaller, at the edges of any cloud. Once you begin to look for cloud iridescence, you will find that it is extremely common and is to be found in many different types of cloud.

If you fly over clouds, you may notice a series of coloured rings around the antisolar point (the point directly opposite the Sun), where the shadow of the aircraft may, or may not, be visible; this is known as a glory. It may also be visible if you are standing above banks of clouds or mist lying in a valley. In these conditions, the glory will be visible only around the shadow of your own head, even though the shadows of any companions will also be visible. They, of course, will experience the same effect.

A glory frequently accompanies the Brocken Spectre, named after the mountain in Germany where it is well known. Here, an apparently magnified shadow of the observer

A corona, *above,* is formed only when a veil of thin clouds covers the Sun. The red tint at the outer edge is nearly always noticeable, but the blue colour closer to the Sun itself is not so frequently seen. This photograph was taken using a wide-angle lens and, as a result, the corona appears to be wider than normal.

▷

Colours in iridescent clouds, *right,* such as a corona, occur only when cloud cover is thin. The tints are normally pale and are rarely as distinct as in this example, photographed over the Alps.

Optical phenomena/5

can be seen on a neighbouring cloud-bank. The glory is also called the Brocken bow.

The precise explanation of a glory is complex since it involves both backward scattering of the light and interference, which together produce a series of diffraction rings.

At times when the sky is covered with a thin, even veil of cloud of the type known as cirrostratus (see pp 46–59), a narrow, pale-coloured ring may be seen surrounding the Sun. This is an example of a halo, which can also occur around the Moon. In contrast to the coronae and similar water-droplet phenomena, the red light is at the inner edge, merging into yellow and, finally, white toward the outside. But this is only one of the halo phenomena that you may be able to see.

Ice crystals have many different shapes and their orientations may be completely random or highly ordered. They may be floating steadily, oscillating slightly, or even rotating. Furthermore, they may reflect light from their flat faces and also act as prisms, refracting and dispersing light that passes through them. The altitude of the Sun also makes a difference.

The most common ice-crystal effects are haloes, the brightly coloured circumzenithal arc (which touches the outer halo), and Mock Suns, or parhelia. The latter may be bright, with white 'tails' extending away from the Sun, and are often seen in isolated patches of cirrus clouds (see pp 46–59). Sometimes a white circle (the parhelic circle) can be seen running parallel with the horizon.

Depending on the extent and thickness of the ice-crystal clouds, only parts of the haloes and arcs may be seen. Movement of the clouds and of the Sun cause the display to change. If you can, watch the display over a long period, scanning all around the sky. Some of the effects are faint and difficult to see, as well as being rare.

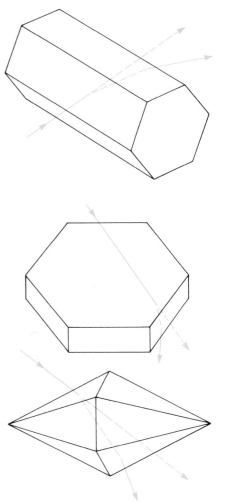

Ice crystals form in many different shapes, a few of which only are shown, *above*. Hexagonal prisms with random orientations give rise to the 22° halo explained *below right*. Flat plates, orientated horizontally, cause mock suns, the circumzenithal arc, and other spectacular effects. Many rare arcs may be produced by prisms with triangular end faces.

A bright, vertically stretched image of the Sun (known as a sun pillar) photographed in Antarctica, *above.*

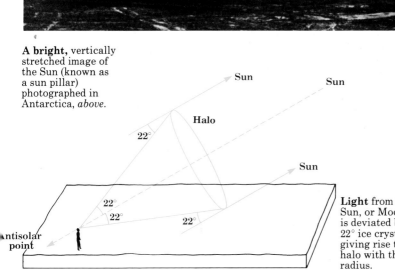

Light from the Sun, or Moon, *left,* is deviated by 22° ice crystals, giving rise to a halo with that radius.

Wind movements/1

The atmosphere is seldom still and so the surface of the Earth is covered by both widespread and localized winds. The major circulation patterns are caused because the equatorial regions receive more heat from the Sun than do the poles. In the process known as convection, warmed air rises and cooled air sinks, forming a circulation in a series of convection cells.

In the atmosphere, warmed air rising over the tropics creates a low-pressure zone at the Earth's surface. At higher levels, it then flows out toward the poles before cooling and sinking once again. At the surface, cooler air, drawn from the middle latitudes, converges toward the Equator; at the poles, cold air spreads out at the surface and returns at higher levels.

If the Earth did not rotate, a single circulation cell would form on each side of the Equator. As it is, three basic cells are formed in each hemisphere – those over the tropics and polar regions circulate in the same direction, while the middle-latitude cell of air circulates in the opposite direction.

The speed at which the Earth's surface rotates is much greater at the Equator than it is at the poles. If you imagine a quantity of air at the Equator, it will be carried eastward at the same speed as the surface beneath it. But if the air moves directly toward one of the poles, its velocity becomes greater than that of the higher-latitude surface areas, and so it is diverted toward the east.

A similar effect occurs with air moving from the poles to the Equator. As a general rule, air is diverted to the right in the northern hemisphere and to the left in the southern. This is known as the Coriolis effect, which changes the north-south flow of air in the three basic cells into the prevailing easterlies and westerlies.

Where the air in the middle-latitude and tropical circulation cells descends

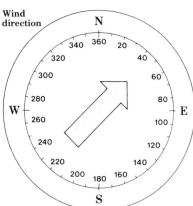

Wind direction

Winds are always described by the direction from which they have come. A south-westerly wind, therefore, blows from southwest to northeast.

▷

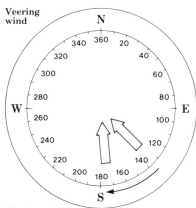

Veering wind

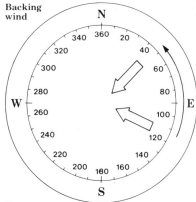

Backing wind

Veering is the term used to describe a clockwise change in wind direction.

Backing describes a counterclockwise change in wind direction.

Cirrus clouds, *top,* sometimes form in distinct, elongated bands. These indicate the location of the jet stream of high-speed winds far overhead.

Wind movements/2

toward the Earth's surface, at approximately 30° North and South, high-pressure regions develop, giving rise to two belts of generally fine and settled weather. The air flowing out of these high-pressure belts produces the dominant trade and westerly winds.

At the boundary between the middle-latitude and polar cells, cold air from the poles tends to undercut the warmer air, giving rise to depressions (see pp 60–1) and unsettled weather. Seasonal variations cause the general pressure pattern to shift north and south; this is more marked in the northern hemisphere because of the disturbing effects of the large land masses.

When pressure patterns are plotted on a chart, lines, known as isobars, are used to connect points with identical pressures. On general isobaric charts, the interval separating the isobars is normally four millibars. Since winds are produced by pressure differences, winds are strong where the isobars are close together.

Pressures vary throughout the atmosphere and, given enough information, a whole series of isobaric charts for different altitudes could be constructed. Most of the pressure readings and other information for the upper air are obtained from radiosondes and weather balloons, which are launched from some meteorological stations. Despite the restricted amount of data available, satisfactory charts can be plotted for the most important levels of the atmosphere.

At the land surface, there is a complicated arrangement of very nearly circular, or elliptical, areas of high and low pressure. These highs and lows become less distinct at greater altitudes, and the resulting normal overall pattern resembles a series of waves. Ridges of warm air point toward the poles and alternate with troughs of cold air stretching down toward the Equator. These pairs of

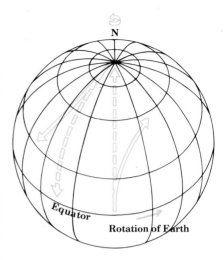

The rotation of the Earth, *above,* causes the mass of air moving toward the poles to be deflected to the right in the northern hemisphere and to the left in the southern hemisphere. This is known as the Coriolis effect.

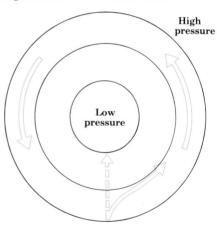

Air tries to flow directly from high pressure to low but is turned aside by the force of the Coriolis effect. When the pressure gradient and Coriolis forces are in balance, wind flows directly along the isobars and around the centres of high- and low-pressure areas.

▷

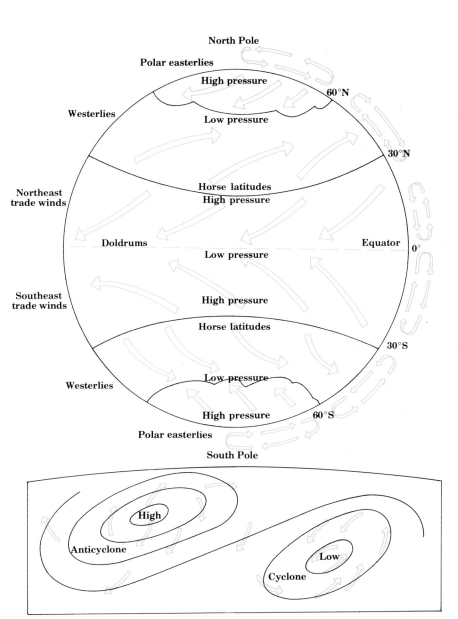

North Pole

Polar easterlies

High pressure

60°N

Westerlies

Low pressure

30°N

Northeast trade winds

Horse latitudes
High pressure

Doldrums

Equator   0°

Low pressure

Southeast trade winds

High pressure

Horse latitudes

30°S

Westerlies

Low pressure

High pressure   60°S

Polar easterlies

South Pole

High

Anticyclone

Low

Cyclone

The global circulation of surface air, *top*, occurs in three main belts – the tropical and polar easterlies and the mid-latitude westerlies. Friction close to the Earth's surface, however, slows the wind, and the pressure gradient reasserts itself. This produces a spiral flow across the isobars out of high-pressure areas and into lows. The northern hemisphere flow is shown, *above*.

Wind movements/3

ridges and troughs are extensive and between four and six only are normally found around the Earth. They tend to move quite slowly and greatly influence the weather at the surface below them.

At high levels, the winds flow along the contours of these waves. Where the pressure gradients are particularly strong, they form what are known as jet streams. Below about 500 to 1,000 m (1,600 to 3,200 ft) surface friction stops the free flow of air, which then tends to spiral outward from high-pressure areas (anticyclones) and in toward the lows (cyclones or depressions). The flow is counter-clockwise around the low-pressure areas in the northern hemisphere and clockwise in the southern hemisphere.

Seasonal variations also produce wind flows. The most important of these are the monsoons that primarily affect the Asian continent. In winter, cold, dry air flows out from the centre of the continent as the northeasterly monsoon. In the late spring this changes to the warm, rain-bearing southwesterly monsoon. Other, less-pronounced monsoons are found in other parts of the world.

A sea breeze is an example of a more localized effect caused by the temperature difference between the land and the sea. During the day the land is warmed more rapidly than the water. Warm air above it tends to rise and flow out to sea at a higher level, while at the surface, cool air streams in from the water. At night, the situation is reversed and a land breeze blows out to sea. Similar effects occur near any large body of water.

In mountainous areas, heating during the day causes mountain winds to flow up the valleys and hill slopes. The opposite occurs at night, when the air cools and flows down the valleys. Winds that flow down from high land are known as katabatic winds, and the

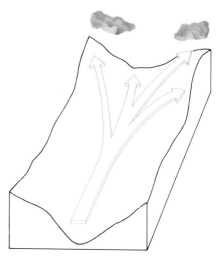

Overnight fog is dispersed by heat from the Sun. Warm air rises up the sides of valleys and may cause clouds to form over the ridges of hills later in the day. The effect of the rising air can produce strong mountain winds blowing up the valleys.

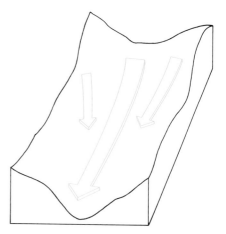

Valley winds consist of air that cools during the night and subsides into the valleys. In narrow valleys there may be a considerable funnelling effect that increases the strength of the wind. After it emerges from the valleys, the cold air may spread out over large areas of the plains below.

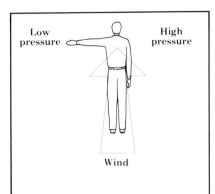

Low pressure / High pressure

Wind

A hand-held anemometer allows a meteorological observer to read wind speed direct. In this design, the rotation of the shaft is transmitted through a magnetic-drag coupling to the wind-speed scale. Other types give a digital read-out.

Buys-Ballot's law can be used to find the approximate direction of a low-pressure centre. If you stand with your back to the wind, the low pressure is to the left (to the right in the southern hemisphere), and the centre is about 30° in front of this.

The Beaufort Scale

Symbol	Force	Description	Result	Wind speed (km/h)
	0	Calm	Smoke rises vertically	0–1
	1	Light air	Wind direction indicated by smoke but not wind vanes	1–5
	2	Light breeze	Wind noticeable, leaves move, wind vane moves	6–11
	3	Gentle breeze	Leaves, small twigs in constant motion	12–19
	4	Moderate breeze	Wind raises dust	20–29
	5	Fresh breeze	Small trees start to sway	30–39
	6	Strong breeze	Large branches in motion, wind whistles	40–50
	7	Near gale	Trees in motion, awkward to walk against wind	51–61
	8	Gale	Twigs break, hard to walk	62–74
	9	Strong gale	Some structural damage may occur, slates removed, etc.	75–87
	10	Storm	Trees uprooted, considerable structural damage	88–101
	11	Violent storm	Widespread damage	102–117
	12	Hurricane	Widespread damage	118+

35

Wind movements/4

strongest occur when pools of air over high, ice- or snow-covered areas become extremely cold and dense and eventually cascade down the valleys. The mistral, which flows down the Rhône valley from the Alps, is probably the most famous example of such a wind.

Wind speeds are measured by instruments known as anemometers; the most common type has a set of cups that spins in the wind and causes a shaft to rotate. This either generates an electrical signal or the rotations are counted to derive the velocity of the wind. Wind speeds are expressed in miles or kilometres per hour or in knots (1 knot=1 international nautical mile per hour, or 1.85 kilometres per hour). When precise measurement is not possible, wind speeds are estimated by using the Beaufort scale, which is often used in public weather forecasts.

Air masses are large volumes of air that originate over certain source regions. As they move across the surface of the Earth, they tend to retain the characteristic features, such as temperature and humidity, with which they formed.

However, the air masses do alter to some extent, changing their temperature and losing or gaining moisture. This can greatly alter their stability and, therefore, the type of clouds formed within them.

The four major classifications of these air masses are: Equatorial (E), Tropical (T), Polar (P), and Arctic (A). The classes are further distinguished as being either continental (c) or maritime (m) in origin (except Equatorial air, which is only maritime in nature). Continental air is generally dry (especially so in winter), and maritime air is humid.

The diversions between air masses are marked by fronts – either warm or cold, depending on the air mass that is advancing, or occluded, when the

Air masses

Arctic (A)
Polar regions; cold and dry
Maritime polar (mP)
Polar regions and long tracks over the sea; cool and moist – lower layers slightly warmer
Continental polar (cP)
Polar regions and long tracks over land; cool and dry, often very dry
Tropical (T)
Subtropical high-pressure regions; warm and dry
Maritime tropical (mT)
Subtropical oceanic regions; warm and moist, often very moist
Continental tropical (cT)
Subtropical land regions; warm and very dry
Equatorial (mE)
Tropical seas; very warm and very moist

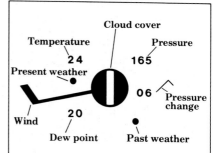

Station plot
This example of a station plot is for Casablanca. Cloud cover is 7/8; pressure 1016.5 mb; pressure change in the last 3 hours 0.6 mb higher, rising then falling; past weather intermittent rain; dew point 20°C; wind southwest by west, 10 knots; present weather intermittent rain; air temperature 24°C.

▷

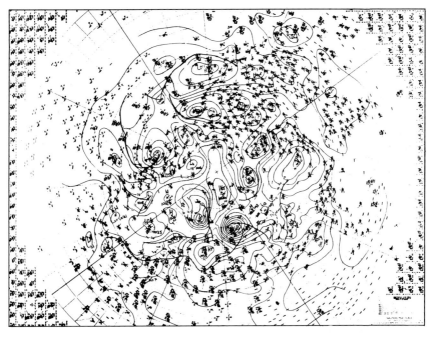

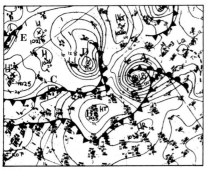

The position of
the fronts in the
map, *top,* can be
seen, *above.* Warm
fronts have semi-
circular cusps and
cold fronts
triangular, while
occluded fronts
have both. The
cusps point
forward, showing
the way the fronts
are moving.

**The weather
map,** *left,* shows
part of a northern
hemisphere
isobaric chart for
12.00 GMT on
28 September 1978.
The deepest
depression for that
day (central
pressure 985 mb)
was centred over
the Atlantic, south
of Iceland (A). The
highest pressure
(1035 mb) lay to its
southwest, east of
Newfoundland (B).
Over the St
Lawrence River a
new depression
was forming (C) on
the trailing cold
front from a well-
occluded low (D)
covering the Davis
Strait. A weak
depression (E) lay
over the border
between the USA

and Canada. Low-
pressure troughs
(F and G) were to
be found over
central Europe and
the western
Atlantic. The
observational data
from each
individual
reporting station
or vessel has been
computer-plotted
in the standard,
international code.

37

Wind movements/5

warm air has been lifted from the surface. Fronts are only rarely stationary, and they are usually accompanied by considerable changes in wind speed and direction. On a weather chart, the isobars frequently display a change in direction at fronts; this is particularly the case with cold fronts.

The main polar fronts, between polar and tropical air, are extremely unstable. Wavelike features continually develop and become full-scale depressions (see pp 60–9), which then travel around the world, generally eastward. These depressions are the major weather features affecting the middle and higher latitudes in both the northern and southern hemispheres.

The production of the many charts that are required for weather forecasting requires measurements from hundreds of observation stations all over the globe. These stations include both merchant ships and special weather ships in mid-ocean, as well as automated monitoring buoys and unmanned land stations in remote regions. A much smaller number of stations make upper-atmosphere observations by balloon-borne radiosondes. These lightweight instruments are also tracked by radar to provide details of the speed and direction of upper-level winds. Aircraft and satellites, too, contribute valuable information.

There is an efficient communications system that passes encoded messages, with details of the individual observations, from all the stations to national centres and then, via a truly international network, to other forecasting offices. The principal coordinating body is the World Meteorological Organization (WMO), which has established standardized descriptions of meteorological phenomena, coding and plotting symbols. These are now used and understood by meteorologists all over the world.

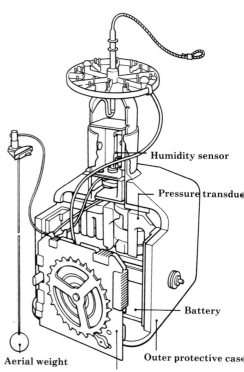

Humidity sensor

Pressure transducer

Battery

Aerial weight

Outer protective case

Printed circuit board

A radiosonde and its instrumentation are designed to be light in weight and, therefore, easily carried aloft by balloon. Pressure, temperature and humidity readings are taken by the various sensors and transmitted to the ground. A battery provides power for the sensors and radio. The balloon also carries a metallic reflector which enables it to be tracked by radar. When the balloon bursts, a parachute returns the instrument package to earth.

A typical temperature profile, *below,* as recorded by a radiosonde, shows a major alteration at the tropopause.

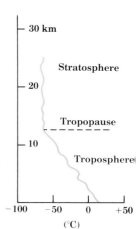

30 km

Stratosphere

20

Tropopause

10

Troposphere

−100   −50   0   +50

(°C)

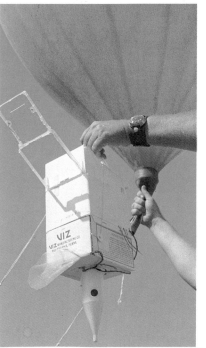

Meteorological balloons are of two types: those that carry radiosonde instrumentation, *left*, and the simple, non-instrumented form of pilot balloons. Both may be inflated with hydrogen or helium. Pilot balloons are inflated so that they rise at a steady rate – 120 m (400 ft) or 150 m (500 ft) per minute – enabling their altitude to be calculated easily.

Hoar frost, *above*, consists of soft, white crystals. It forms when the water vapour in still, humid air freezes directly on trees and other objects. Rime is similar, but grows on upwind surfaces when super-cooled fog flows past objects on the ground.

Water in the atmosphere/1

Nearly all weather phenomena are caused by water in one form or another: fog, clouds, rain and snow are obvious examples. Although water is a common substance, it is unique in that it occurs regularly in the atmosphere in its three different forms – liquid, solid (ice) and gas (water vapour).

Most water vapour arises through evaporation from the sea, although there is a smaller contribution from the land. Evaporation requires considerable heat, since the molecules of water must gain sufficient energy to enable them to escape from the surface of the liquid. It takes about 600 calories of energy to evaporate 1 gram of water. This same amount of heat (known as latent heat) is released when water vapour changes back into liquid form.

The amount of water vapour contained in the atmosphere is variable. On occasions it may be as high as 4 per cent by volume, while at other times the air may be almost completely dry. As the temperature of a particular volume of air increases, so does the total amount of water it can hold, until saturation point is reached. Most of the time, however, it holds only a fraction of the water it could contain, and this fraction, expressed as a percentage, is known as the relative humidity.

High relative humidities are generally uncomfortable; in summer when the temperature and humidity are high, the weather feels oppressive. In winter, damp air feels raw and cold, whereas, even at low temperatures, dry air remains bracing.

One way to measure humidity is to use a hygrometer. This instrument contains a small bundle of fibres – frequently human hair – that lengthens when the humidity is high and contracts under dryer conditions. The electrical conductivity of certain chemicals also varies with humidity, and this principle is used in some

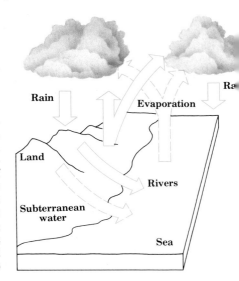

Rain

Evaporation

Ra

Land

Rivers

Subterranean water

Sea

Most of the water in the atmosphere exists as a result of evaporation from the oceans by the heat of the Sun. A small amount comes from the land. Eventually, the water returns as precipitation (rain, snow or hail). About three-quarters returns to the sea; the rest falls on the land, so completing the water cycle.

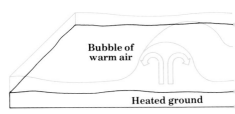

Bubble of warm air

Heated ground

Convective bubbles of air form when a layer of heated air at the surface breaks away and rises as a thermal. If it rises far enough to reach the condensation level before it disperses, a typical small cumulus cloud is formed.

▷

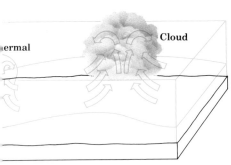

Cumulus clouds of different sizes, *top*, indicate considerable convection and the possibility of showers later in the day.

Hot air rising from a fire, *right,* has broken away as a thermal with an independent existence. A small cumulus cloud is forming above the trails of smoke.

Water in the atmosphere/2

instruments, especially those carried aloft by radiosondes (see pp 30–9).

The simplest way to measure humidity is to use a pair of thermometers, one of which has its bulb covered with a wick moistened with water. The evaporation of water removes energy from the thermometer, so the temperature drops. The temperature difference between the wet and the dry bulb thermometers can then be used to calculate the relative humidity.

If unsaturated air is cooled, its relative humidity increases until eventually it reaches 100 per cent and becomes fully saturated. Any further cooling causes the water vapour to condense into droplets. There must, however, be a surface of some sort on which the water can condense. In the atmosphere, microscopic dust, salt and smoke particles exist in unimaginable numbers, and these act as condensation nuclei.

When air is compressed its temperature rises and, conversely, when it expands it becomes cooler. In the atmosphere, any bubble of air that rises moves into a region of low pressure, which causes it to expand and become cooler. Rising, unsaturated air cools at 1° C per 100 metres (328 ft) and, if it continues its ascent, it eventually reaches saturation.

Although under normal conditions water on the ground freezes at 0° C (32° F), cloud droplets can exist unfrozen at much lower temperatures, certainly as low as −40° C (−40° F). Such droplets are said to be supercooled, and they freeze rapidly on impact with any surface or in the presence of suitable nuclei. For example, many successful rainmaking experiments have been carried out, some of which have relied upon seeding supercooled clouds with silver iodide crystals, which closely resemble ice crystals in their structure. The ice crystals grow rapidly and later melt and fall out of the cloud as rain.

Banner cloud

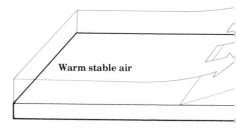

Warm stable air

Uplift of air commonly occurs at fronts, where warm and cold air meet. When the rising warm air is stable, *above,* the clouds are often restricted in depth. The associated rain is light, but sometimes prolonged. When the air is unstable, *below,* there can be enormous upward growth of clouds. Such towering clouds, frequently found at cold fronts, can give rise to heavy rain and the chance of thunderstorms.

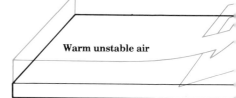

Warm unstable air

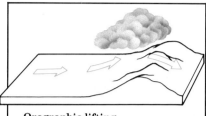

Orographic lifting
This movement occurs when warm, moist air is carried over a mountain range. This is shown by banner cloud, *left,* which has formed over a peak in the Andes. Depending upon temperature and the level of the dewpoint, cloud may shroud the tops of hills or lie some way above them.

Stratus clouds

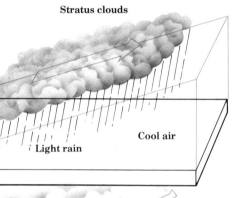

Cool air

Light rain

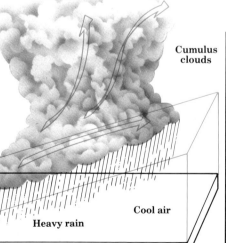

Cumulus clouds

Cool air

Heavy rain

Stable and unstable air
A bubble of unstable air continues to rise only if its temperature is warmer than that of its surroundings. If the surrounding air temperature is lower, then the bubble remains unstable and continues to rise. Stable conditions occur if the surrounding air is warmer. Then the air in the bubble is denser than its surroundings and so it tends to sink back toward the ground. Similar effects occur if air is forced downward and warmed by compression. Stable air attempts to return to its original level, while unstable air continues to sink.

Cloud watching

One of the joys of watching the day-time sky is observing the clouds. They appear in so many different shapes and sizes that identification seems a hopeless task, especially when many forms are present at the same time. But with a little practice it becomes much easier, especially if you can begin by identifying some of the more distinct types. Under certain circumstances, too, such as at the various fronts of depressions, clouds tend to follow each other in certain sequences, which can help with identification.

Clouds may be broadly divided into three forms – heaped clouds, layered clouds, and clouds with a feathery or fibrous appearance. These are known as cumulus, stratus and cirrus, respectively. Except for the addition of some subdivisions, these cloud classifications have been in use since the early nineteenth century. As with other aspects of meteorology, the World Meteorological Organization has standardized cloud descriptions, so that reports, abbreviations and symbols are the same all over the world.

Part of cloud classification is based upon height, and this is not always easy to judge. Clouds can be regarded as occurring at three different levels in the troposphere (see pp 10–13), and thus are known as low, medium and high clouds. Low clouds exist in the bottom 2 to 3 km (1 to 2 mls) of the atmosphere. Above this come the medium clouds, which may reach heights of about 4 km ($2\frac{1}{2}$ mls) close to the poles or 8 km (5 mls) near the Equator. The high clouds are found right up to the tropopause.

Some cloud types tend to occur in one or other of these layers, but this does not always happen, since temperature and water content also have an effect. In addition, some clouds may be extremely deep; for example, cumulonimbus clouds, may extend from the lowest level to the uppermost levels of the troposphere.

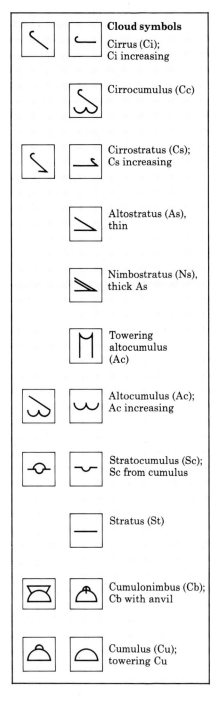

Cloud symbols

Cirrus (Ci); Ci increasing

Cirrocumulus (Cc)

Cirrostratus (Cs); Cs increasing

Altostratus (As), thin

Nimbostratus (Ns), thick As

Towering altocumulus (Ac)

Altocumulus (Ac); Ac increasing

Stratocumulus (Sc); Sc from cumulus

Stratus (St)

Cumulonimbus (Cb); Cb with anvil

Cumulus (Cu); towering Cu

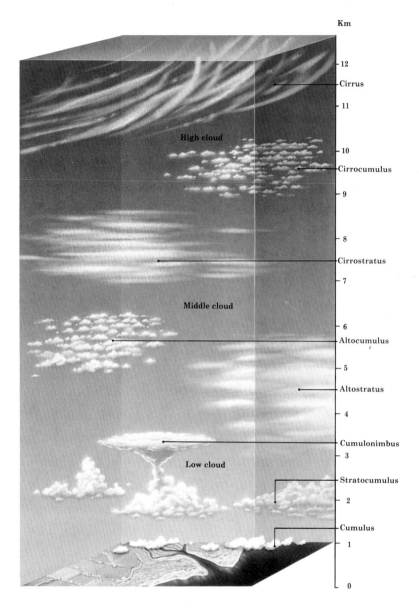

Km
— 12
Cirrus
— 11

High cloud
— 10
Cirrocumulus
— 9

— 8
Cirrostratus
— 7

Middle cloud
— 6
Altocumulus
— 5
Altostratus
— 4

Cumulonimbus
— 3
Low cloud
Stratocumulus
— 2

Cumulus
— 1

— 0

Clouds occur in three fairly distinct zones. The high cirrus clouds consist of ice crystals. Altocumulus and altostratus are found in the middle layer. Close to the ground stratus forms fog or low-lying cloud. Stratocumulus may be formed by the spreading out of cumulus clouds.

45

Cloud varieties/1

The day often starts with the sky quite clear, but as the warmth of the sunlight increases, small fluffy heaps of cloud begin to appear. These are typical cumulus clouds (Cu) that form at the tops of thermals, which rise from the heated ground, become saturated and condense into water droplets.

The first small clouds often appear ragged, with irregular bases, but as the clouds become larger the flat bases that mark the condensation level are easy to see. Since the clouds are usually widely separated, perspective often causes the bases – although all at the same level – to appear as a series of steps down toward the horizon.

Many early-morning thermals are weak and die away, but as the power of the Sun increases during the day, the clouds become larger and more persistent, and they may spread to cover a large area of the sky. This will prevent much of the ground from being heated by the Sun, so restricting the clouds' further growth.

Cumulus clouds also arise over the sea when cool air passes over warmer water, and they may be carried inland for considerable distances by the wind. These particular clouds are often more persistent since there is less temperature difference between day and night.

Large cumulus clouds consist of many convective bubbles of rising air. If you scrutinize their rounded, cauliflower-like tops (binoculars are useful for this – see pp 192–5), you can see the individual towers growing upward, expanding, and then gradually losing their shape. The tops of all types of cumulus appear white where they are illuminated by the Sun; their bases are grey and may become dark as the clouds grow larger and thicker.

The final size and vertical extent of cumulus clouds depend upon the stability of the higher layers of air. Sometimes the vertical growth is restricted to give small, flattened 'pancakes' of

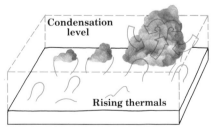

The growth of cumulus clouds is controlled by various factors.

Weak thermals produce small clouds, *left*. With a great difference in

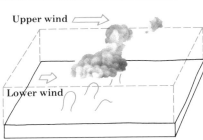

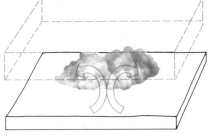

wind speed, *centre*, cloud towers are high. If a stable layer lies just above the condensation level, *right*, upward growth is arrested. **Cumulonimbus** clouds may produce intense, localized showers. The heavy rain is clearly visible in the photograph, *top*.

Cloud varieties/2

cloud. Generally, however, cumulus clouds are associated with instability, so there is considerable vertical growth. Commonly, rising cells of air grow upward together and produce a wide tower of cloud.

Even more extensive and higher than towering cumulus are the giant cumulonimbus clouds (Cb). These are striking, mountainous clouds with ominously dark bases. Although their tops often seem remarkably solid and brilliantly white against the blue of the sky, look for the wispy, fibrous edges beginning to form on the highest cells. These show that some of the water is freezing into ice crystals. This, and the heavy grey bases, indicate that rain will occur before long, even if it is not already falling. This type of cumulus may develop into a full-scale thunderstorm.

In complete contrast to the family of cumulus clouds, stratus, or layer, clouds are associated with stable atmospheric conditions. They are also different in that they are restricted in their vertical depth. Even though some varieties may, at times, be dense enough to cut off most of the sunlight, giving rise to gloomy days, they are generally extremely thin. What they lack in thickness, stratus clouds make up in horizontal extent. They can form vast sheets of cloud and it is not unknown for them to extend horizontally for more than 1,000 km (620 mls).

In addition, such widespread layers of cloud can only arise under stable conditions. Stratiform clouds are generally caused by gentle uplift, either over another layer of cooler air or by slowly rising ground. Unlike cumulus clouds, no convection takes place in stratiform clouds, although turbulence caused by the ground may itself promote the formation of stratus (St). Thus, the lowest of these layer clouds frequently has a base well below 200 m (650 ft), so that it obscures the tops of nearby hills or tall buildings. If you

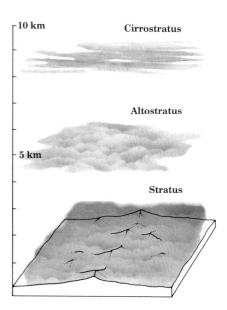

Ordinary stratus forms on, or just above, the surface of the Earth. The altitudes of altostratus and cirrostratus vary with conditions, but typical heights are indicated here. Altostratus consist largely of water droplets, while cirrostratus contains ice crystals.

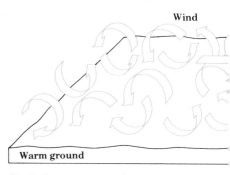

Typical conditions for the formation of stratus cloud occur when a layer of warm, moist air, which has formed above a warm surface is carried over colder ground or water. High winds may produce too much turbulence and

▷

A layer of stratus cloud covers the tops of the hills, *above.* In the colder layers, higher in the atmosphere, cirrus clouds merge into cirrostratus.

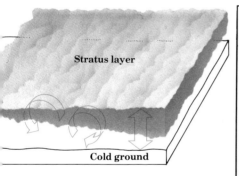

Stratus layer

Cold ground

insufficient cooling for cloud to form, unless the difference between air and ground temperatures is extremely large.

Fog
Two main processes are involved in fog formation. One is identical to that which produces stratus but with lighter winds and stronger cooling. Such fog may be persistent in coastal areas but may disperse more quickly over the land as the air warms up during the day. The second process takes place on clear nights, when the lower air layers radiate their heat away into space. It may take several hours for the dew-point to be reached, but cooling accelerates rapidly after the first droplets have condensed. Sunlight or an increase in wind speed lifts and disperses the fog entirely.

Cloud varieties/3

are in such a building or on high ground, the cloud is indistinguishable from fog. Even if the layer of cloud is not widespread, the uplift provided by quite low hills may still be sufficient to cause their tops to be shrouded in grey cloud.

Because stratus is thin, you can often see the outline of the Sun through this type of cloud. Even though the cloud consists of water droplets (see pp 40–3), it does not show any form of corona. When the layer is broken into patches, stratus cloud is ragged and has no distinct edges.

Although stratus may be damp and unpleasant, it does not produce much rain, only a gentle drizzle. If the conditions are cold enough, there may be some small snow or ice crystals. In this, it is quite unlike the related, but much thicker, nimbostratus cloud, which is associated with depressions and may give heavy and prolonged periods of rain.

A similar type of cloud at the middle layers is altostratus (As). This may also vary in thickness, and, since it is higher than stratus, these variations are more obvious. Altostratus is distinguishable by the gradation of colour, which ranges from fairly dark grey, where the cloud is thick or in shadow, to a much lighter, bluish tint. It is often possible to see the Sun through the thinner patches, when it appears diffuse, as if it were shining through ground glass. Altostratus may cover the sky or occur only in isolated patches, when it frequently has distinct edges.

Altostratus clouds are not particularly exciting to observe, but they are one of the indicators of the approach of a depression as they gradually thicken and merge into heavy rain-bearing nimbostratus. Thick altostratus may itself occasionally give rise to rain, although this often fails to reach the ground.

Even higher than altostratus clouds

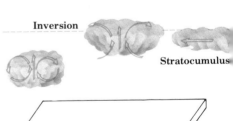

Inversion

Stratocumulus

Stratocumulus cloud forms when an inversion acts as a 'lid' on convection, *above*. Thermals rise freely until they reach the inversion, which they may overshoot slightly. The individual clouds that are formed spread sideways. They may join to give a continuous layer of stratocumulus.

▷

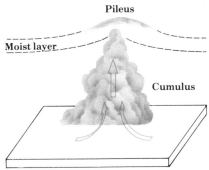

Pileus

Moist layer

Cumulus

Altocumulus clouds appear in different forms, but always exhibit light and dark shading, *top*. They often arise when a sheet of altostratus breaks up into patches.

Pileus cloud often forms when nearly saturated air is uplifted by a rsing cumulus cell, *above*.

The resulting pileus cloud, *right,* is often broken by the still rising cumulus.

Cloud varieties/4

—and usually preceding them ahead of an approaching depression – is cirrostratus (Cs). This is a thin, transparent ice-crystal cloud through which the Sun is always visible. It may occur as a complete, continuous veil stretching from horizon to horizon, or it may allow broken patches of blue sky to show through in places.

Although often a smooth, apparently featureless sheet, cirrostratus may sometimes show a distinct fibrous structure. It is closely related to cirrus, which often thickens into cirrostratus. The cloud-sheet can be thin enough to be almost invisible. However, if the sunlight is slightly weaker than you might expect, and the sky is slightly milky, it could well be that a thin sheet of cirrostratus has spread over the sky. These clouds consist of ice crystals and are ideal for the formation of halo phenomena.

Corresponding to the three main stratiform clouds are the three varieties that combine features of both layer and heap clouds. Here, a general layer of clouds is broken into cumulus-like elements, which may be fairly isolated tufts, 'pancakes', rounded masses or long, distinct rolls.

Stratocumulus (Sc) is the lowest stratiform variety and is common. It is often produced when active cumulus clouds grow upward until they reach a stable layer, when atmospheric inversion causes the temperature to increase with height, instead of decreasing. The cloud cells spread out sideways and may produce an unbroken blanket of cloud. Because it consists of a large number of cloud elements, stratocumulus appears as uneven shades of grey.

Altocumulus clouds (Ac) give some of the most attractive skies. They sometimes occur as evenly spaced tufts, looking like balls of cotton, and at others as regularly arranged, long rolls of cloud. They are much lighter in colour than stratocumulus, but the

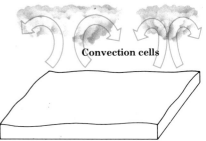

Convection cells

Thin stratiform clouds break up into cumulus types when shallow convection cells form within the cloud layer, usually because the upper surface loses heat by radiation. In general, the thicker the original layer of cloud, the larger the individual cloudlets.

▷

Altocumulus clouds, together with individual patches of altostratus (especially in the distance) can be seen, *above.* Precipitation is falling from the clouds, producing virga trails in the slower-moving air.

Cirrocumulus clouds, *right,* have a delicate structure. The cloudlets are always smaller than altocumulus and are without distinct shading.

Cloud varieties/5

individual elements nearly always show darker shading at their bases. Altocumulus clouds are composed of water droplets and show some of the most attractive coronae and iridescent effects.

Higher still come the cirrocumulus clouds (Cc). These also frequently show a range of regular structures. They are so high, however, that the individual elements look small from the ground, and the layer may have a delicately rippled appearance. The clouds are thin and white, and the elements show no signs of darker shading. Temperatures at these levels are low, so cirrocumulus clouds consist of either supercooled water droplets (see p 40–3) or ice crystals.

All these three cloud types – Sc, Ac, and Cc – may show billows or regularly spaced, parallel bands of clouds, normally approximately at right angles to their motion across the sky. These bands arise when there are two layers of air moving at different speeds. This wind shear produces a series of waves in the sheet between the layers and cloud forms at the crests of the waves.

A somewhat similar effect occurs when an air stream is forced to cross a range of hills. This sets up a series of lee waves, which may extend for a considerable distance downwind. Clouds form in the crests of the waves and remain stationary with respect to the hills – unless there is a change in the wind speed or direction. These mountain waves are often large and reach high into the atmosphere.

Mountain waves can be used by glider pilots to gain great altitude without venturing into dangerously violent cumulonimbus clouds. Stratocumulus, altocumulus and cirrocumulus may all form wave clouds, but the two higher varieties are often conspicuous from the ground.

There is one other highly distinctive variety of cloud – cirrus (Ci). This

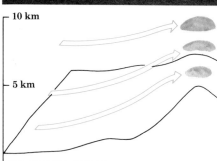

Wave clouds, or altocumulus lenticularis, *top*, are aptly described by the French term *piles d'assiettes* (piles of plates). Each plate is formed by a single layer of humid air.

▷

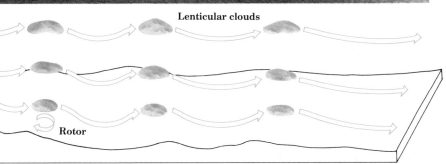

Lenticular clouds

Rotor

Wavelike motions can be set up when air rises over mountains, *above*. Lenticular clouds form at the top of each wave. Close to the ground, an eddy may develop into a strong circulation known as a rotor.

55

Cloud varieties/6

cloud forms delicate, feathery wisps, hairlike streaks and tufts high in the sky. It occurs also in long bundles that stretch for great distances, justifying its common name of 'mares' tails'.

Cirrus clouds consist of ice crystals that have formed in small generating heads of cloud and then fallen down to lower levels. When the layers of air are moving at about the same speed, cirrus appears as twisted tufts. More frequently, however, the lower layer is slower moving, so the crystals lag far behind the actual heads of the trails, which can be many kilometres long. If you see the sky gradually becoming covered with streaks of cirrus that finally merge into a sheet of cirrostratus, you can be fairly certain that a depression and accompanying bad weather are approaching.

Extremely active cumulonimbus clouds often reach great heights, where large numbers of ice crystals are formed in their upper layers. You can see this happening when the tops of the clouds develop a fibrous or feathery appearance. If the conditions are right, high-level winds may draw out the crystals into cirrus-like streaks that form a massive, overhanging anvil cloud. The top of this anvil is often at the level of the tropopause itself, where the increase in temperature causes an inversion and stable conditions, which prevent the cells of cumulonimbus from rising much higher. Depending on the wind speed, the edge of the anvil may be 20 km (12 mls) or more from the parent convection cells within the cumulonimbus cloud.

When cumulonimbus clouds develop feathery, ice-crystal tops, you can be sure that showers of rain will soon begin to fall. Normal cloud droplets are so small that they remain suspended in the air, and their motions are controlled by up- and downdraughts within the cloud. Ice crystals, however, are much heavier,

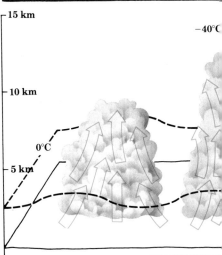

15 km

−40°C

10 km

0°C

5 km

The life of a shower cloud. *top,* is traced in the diagram, *above.* In the initial stages, when the cloud is small, it contains mainly raindrops, and surface rain is slight. In the mature stage, ice pellets may be present in the cloud. Downdraughts may become strong

▷

Slight	Continuous	Intermittent moderate at observation time	Continuous moderate at observation time	Intermittent heavy at observation time	Continuous heavy at observation time
(drizzle)					
(rain)					
(snow)					

The symbols for the most common forms of precipitation are illustrated, *above*: top, drizzle; centre, rain; bottom, snow. These symbols are standard and are used by meteorologists throughout the world when reporting weather conditions.

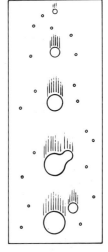

and rain heavy. At a later stage, there are few updraughts, the rain becomes light and the cloud begins to decay.

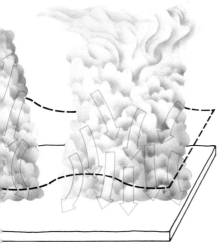

Raindrops can form in two ways. Sometimes water vapour may freeze directly on ice crystals, *left*. These grow as they fall and eventually melt into rain. In warm clouds, a few large initial drops collide with smaller ones, *right*. Beyond a certain size the drops split.

Cloud varieties/7

and they grow and fall rapidly, eventually melting into raindrops when they reach the lower, warmer layers of air.

Some clouds never reach freezing point, even in the highest layers, yet still give rise to heavy rain, especially in the tropics. It now appears that these clouds must contain particular condensation nuclei – especially salt – that encourage large droplets to form. These are sufficiently heavy to be able to fall out of the cloud as rain. Salt particles in the atmosphere have normally come from the sea, but rain-making experiments have shown that it is possible to produce rainfall by seeding warm clouds with common salt. Spraying clouds with large droplets of water can also act as a trigger.

Apart from cumulonimbus 'shower' clouds, the main rain-bearing clouds are nimbostratus (Ns). This is a heavy grey layer of cloud, which is always thick enough to obscure the Sun completely. Unlike most of the other clouds, it has no well-defined base level, since there are more or less continuously falling curtains of heavy rain or snow.

There is no guarantee that raindrops or ice crystals from clouds will reach the ground. The layers below the parent cloud are unsaturated, so the rain evaporates as it falls. Depending on temperature, ice crystals may melt, or else pass directly into water vapour in the process known as sublimation.

Falling ice crystals in a cloud may encounter droplets which freeze on them, gradually building up into snowflakes that float downward. If the temperature at the Earth's surface is below 4° C (39° F), these flakes will reach the ground; but they will immediately begin to melt unless the temperature is even lower. Raindrops falling into a low-temperature layer sometimes freeze into ice pellets, known as sleet; hail is formed by a different process (see p 66).

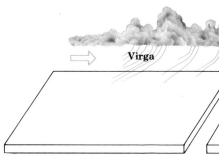

Typical wispy cirrus clouds were photographed over Antarctica, *top*. The streaks tend to thicken into cirrostratus.

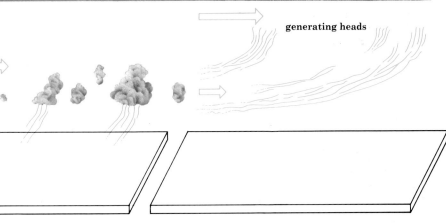

generating heads

Fallstreaks (virga) may arise from several varieties of cloud, such as towering altocumulus, *left,* and broken altocumulus, *centre.* The generating heads of cirrus, *right,* are rarely seen.

Nature's fireworks/1

In meteorology a depression is known as a wave cyclone. The word 'cyclone' describes the circulation and pressure patterns and must not be confused with the violent tropical storms. These storms should correctly be referred to as tropical cyclones.

Most depressions arise from fluctuations, or waves, at the polar front, which separates warm, subtropical air from cold, polar air. Although fronts and depressions may form at the boundaries between other air masses, disturbances at the polar front are the most important.

A front forms a sloping boundary between two air masses with different densities and temperatures. The front slopes away from the region of warm air, which is always above the colder air. At a warm front, where warm air is advancing, the front slopes at a shallow angle, typically about 1 in 150.

As the warm air rises, it becomes saturated, causing clouds. The approach of a depression is first indicated by high cirrus clouds, which slowly increase until they cover the sky. Aircraft condensation trails (contrails) tend to persist and do not evaporate as they would in drier air. This is a strong indication that a depression and bad weather are on the way. The streaks of cirrus show the direction of the upper wind and of the approaching front. As the cirrus turns into a gradually thickening sheet of cirrostratus, haloes are more likely to be seen.

The second type of cloud warning of an approaching front is altostratus, which thickens until the sun is not visible. As the ground cools down, convection diminishes and any lower cumulus clouds begin to disperse. Eventually the altostratus turns to nimbostratus and rain or snow begins to fall.

With the arrival of the warm air, the rain lessens or ceases entirely, and wind at the Earth's surface will usually veer (see pp 30–1) more toward

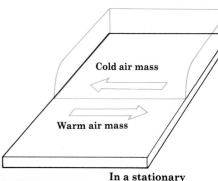

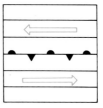

In a stationary polar front, the cold air lies to the north (in the northern hemisphere). On an isobaric chart, *left,* the winds are visualized as flowing smoothly along the isobars. The symbols indicate the direction in which the air masses are moving.

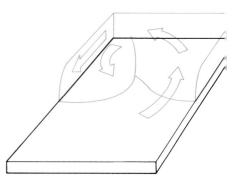

A fully developed depression can now be seen. The low pressure has deepened and the warm sector is fully developed between the fronts. The cold front is rapidly overtaking the warm front.

▷

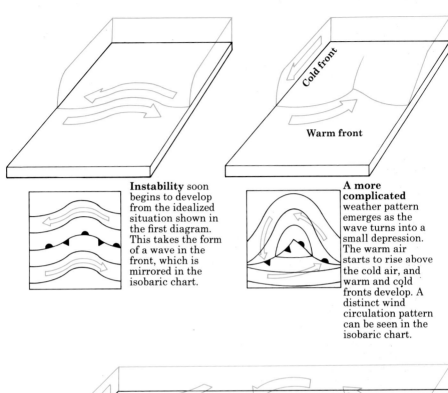

Cold front

Warm front

Instability soon begins to develop from the idealized situation shown in the first diagram. This takes the form of a wave in the front, which is mirrored in the isobaric chart.

A more complicated weather pattern emerges as the wave turns into a small depression. The warm air starts to rise above the cold air, and warm and cold fronts develop. A distinct wind circulation pattern can be seen in the isobaric chart.

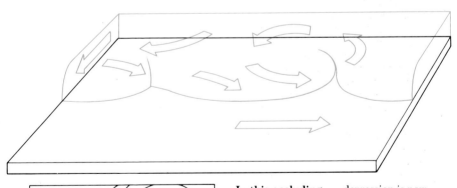

In this occluding depression, the cold front has overtaken the warm and has lifted a pool of warm air away from the Earth's surface. The old depression is now beginning to die away. However, a small secondary wave is developing on the trailing cold front, and this may turn into a second depression.

Nature's fireworks/2

the west. Except near a low-pressure centre, the clouds tend to break up after the warm front passes, leaving patches of blue sky and, sometimes, a mixture of clouds at various levels.

Cold fronts are about twice as steep as warm fronts, often sloping at about 1 in 75, and the clouds and rain generally arrive with and after the front itself. When compared with the conditions at warm fronts, the deterioration in the weather is much more rapid and violent, but the rainbelt is only about half the width. Cold fronts advance rapidly, and sometimes lobes of high, cold air may lie ahead of the surface front, producing great instability (see p 43)

A cold front, therefore, often has a long band of towering cumulonimbus clouds accompanied by heavy rain. It is followed by clearer skies with individual convective clouds. With the passing of such a front, there is a great difference in the feeling of the air, which becomes colder and more bracing. The wind, too, veers more sharply than at warm fronts, usually turning northwest.

Unfortunately for skywatchers, warm and cold fronts only occasionally show these classical forms. They are greatly affected by the characteristics of the air masses and the conditions of the upper air, all of which, in temperate climates, tend to vary with the seasons.

Both warm and cold fronts may be subdued, being marked largely by thick stratocumulus cloud and light rain. Ahead of such a warm front, cumulus may give way to stratocumulus, with sheets of stratus and thinner stratocumulus between the two fronts. Changes at the fronts are much less marked. On the other hand, when the invading air is unstable warm fronts may be much sharper than normal. The cirrus may then be followed by cirrocumulus, altocumulus and, finally, cumulonimbus.

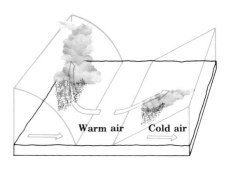

A warm sector of a depression exists when warm and cold fronts are well separated. The sheets of cloud ahead of the warm front may sometimes persist throughout the warm sector.

The circular sweep of cloud viewed from space, *opposite*, marks a depression. The cold air mass, showing broken patches and lines of cloud, lies in the foreground.

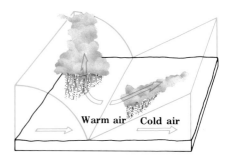

A cold front is more unstable than a warm front, so the associated rainbelt tends to be narrower and heavier. It also moves more

rapidly than a warm front and eventually 'pinches out' the warm sector, giving rise to a continuous belt of rain.

▷

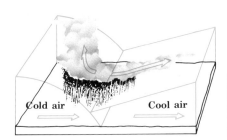

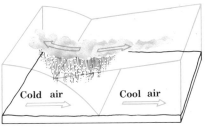

An occluded front forms when warm air is lifted away from the Earth's surface by the cold air. As a result the activity within the rainbelt becomes less vigorous, and rainfall lessens.

A dying depression contains a high pool of warm air and some cloud cover, which may produce a little rain. In a cold occlusion, the second air mass is colder than the first. Sometimes the colder air is ahead of the surface front, which slopes backward rather than forward, producing what is known as a warm occlusion.

Nature's fireworks/3

Vigorous cumulonimbus clouds can develop into full-scale thunderstorms at any time of the year. In winter, such storms do not usually last long or produce more than a few flashes of lightning. They may be more persistent when associated with strong, cold fronts. In summer, conditions favour the formation of towering cumulonimbus clouds with great internal turbulence.

Although individual thunderstorm cells do not persist for more than about 30 to 45 minutes, a decaying cell can trigger the development of another; so a succession of cells can give the impression of a single storm lasting many hours.

Strong heating during the day and considerable instability are important factors in producing thunderstorms. If altocumulus forms vertical towers rather than the normal horizontal formations, this indicates instability, and thunderstorms may develop later.

It is well established that lightning is an electrical discharge between positive and negative regions, and it seems possible that there is more than one process to account for this phenomenon. One suggestion is that as snow pellets freeze they break up, the tiny, positively charged, outer fragments being lifted into the upper regions of the cloud. The heavier cores then carry a negative charge down to the bottom of the cloud.

Whatever the mechanism, positive charges usually accumulate at the top of the cloud and negative ones at the bottom (although small pockets of positive charge do occur at the lower levels). The negative charge in the cloud induces a positive charge on the ground below. This charge 'shadow' sweeps over the ground, keeping beneath the cloud as it travels downwind.

Enormous electrical potentials may be produced, both within the cloud and between the cloud and the ground.

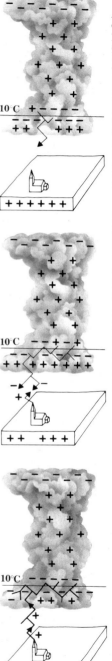

A lightning discharge takes place in several stages. All lightning is a discharge of electricity, which occurs when an electrical field is formed between the upper positive and lower negative charges within a cloud. Negative charges concentrate at the 10°C (50°F) level.

The first branch of lightning, known as the leader, is relatively weak and works down from the cloud along a branching path. When one branch reaches a tall object on the Earth's surface, the negative charge in the cloud drains into the positively charged ground.

A return stroke flashes up to the cloud and then further strokes drain the charge from more distant parts of the cloud. The main lightning stroke is extremely rapid, moving between cloud and ground at a speed that may reach 150,000 km/sec (93,000 mls/sec). It is always the highest point on the ground that attracts the lightning strokes, as can be seen in the photograph, *opposite*.

▷

65

Nature's fireworks/4

These may amount to hundreds of millions of volts, and eventually the electrical resistance of the air breaks down and a lightning flash occurs. This may be between different parts of one cloud, between two clouds, or between the cloud and the ground. The popular view that fork and sheet lightning exist as different entities is incorrect. So-called sheet lightning occurs when the discharge is hidden within the cloud.

The expansion of a column of air that has been violently heated by the discharge causes thunder. The lightning channel may be 2 to 3 km (about $1\frac{1}{2}$ mls) long, so the sound does not reach an observer's ears all at once but produces a prolonged rumble.

The air currents in cumulonimbus clouds are violent, and even when lightning is not produced, the conditions may be favourable for the formation of hail. These hard pellets of ice grow by the accumulation of layers as they are carried upward and downward within the cloud. They are often 5 to 10 mm (less than $\frac{1}{2}$ in) in diameter, but occasionally much larger – the record is 140 mm ($5\frac{1}{2}$ in).

Intense thunderstorms may sometimes form in narrow belts, known as squall lines, that extend across whole countries. These are often associated with cold fronts, but they may occur within a warm air mass. Strong downdraughts from cumulonimbus clouds along a front surge forward ahead of the front, undercutting the warmer air and forcing it to ascend – thus producing the clouds of the squall line.

These may be vigorous and create new storm cells as they themselves die away. At times, a squall line may be 200 to 300 km (125 to 185 mls) ahead of its parent cold front.

Tornadoes are highly destructive, localized phenomena that sometimes form in thunderstorms on cold fronts or in squall lines. In unstable conditions, the violent updraughts found

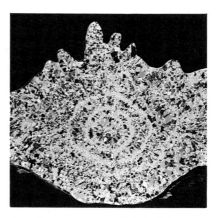

A hailstone, sectioned and viewed in polarized light, clearly reveals its individual layers. The layers were formed from ice of slightly different densities while the hailstone made several circuits within the cloud. It then fell to the ground.

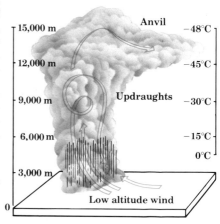

15,000 m   Anvil   −48°C

12,000 m   −45°C

9,000 m   Updraughts   −30°C

6,000 m   −15°C

   0°C

3,000 m

Low altitude wind

0

Hail formation takes place inside the cumulonimbus (Cb) cloud, *above.* Small, initial ice crystals may rise, supported by violent updraughts, and fall within a mature Cb cloud. This process causes the crystals to grow continuously until they finally escape as large hailstones.

▷

About ten lightning flashes illuminated the inside of this cloud as it was photographed. The picture, taken in Hawaii, demanded a 30-sec exposure time. The cumulonimbus clouds rise far above a lower layer of stratiform cloud. Their anvils are at the level of the tropopause. An old disused observatory dome can be seen on the closer of the two extinct volcanic cones.

Nature's fireworks/5

in cumulonimbus clouds can give rise to a narrow low-pressure centre extending below the cloud base. The exceptionally low pressures produce the intense whirling winds found in a tornado. Although direct measurements have never been made, reasonable estimates suggest that wind speeds may sometimes exceed 600 km (370 mls) per hour.

The dramatic drop in pressure is usually sufficient to cause the inflowing air to condense and fill the column with water vapour, producing the pale to dark grey funnel.

Tornado funnels are variable: they may be more or less vertical or snake horizontally for long distances. Their overall lengths range from less than a hundred metres (330 ft) to several kilometres, and diameters can range between 100 and 600 metres (330 and 1,970 ft). The average life of a tornado is 20 to 30 minutes, but some have persisted for several hours.

Tornadoes generally travel across country at about 40 km/h (25 mls/h), although some have been recorded travelling at 100 km/h (62 mls/h) – or even remaining stationary. The funnels do not always reach the ground, but it is common for them to touch down several times along the track, causing intermittent devastation.

Waterspouts are similar to tornadoes, but, as the name implies, they occur over water. They appear to be slightly less powerful and have lesser wind speeds.

A completely different phenomenon is found in the various whirlwinds, which have many local names such as sand devils, snow devils and dust whirls. Localized heating of the ground produces a whirling column of air that may lift loose material from the surface. Whirlwinds normally have lives of a few minutes only, but when the heating is intense, some may become as destructive as miniature tornadoes.

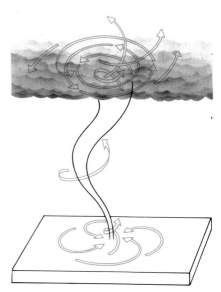

The visible funnel of a tornado, *above*, is the lower part of the vortex. The vortex extends to extreme heights inside the parent cumulonimbus cloud.

Hurricanes cause damage not only with their winds but also through flooding produced by torrential rain. The scene, *above*, shows hurricane destruction in southern Texas.

Tornado funnels usually occur singly, as in the photograph, *opposite*. Further funnels are common, although not all may reach the ground.

The Sun/1

The study of the Sun is a fascinating activity, but it can be dangerous unless you take proper precautions (see pp 182–3). Never be tempted to look directly at the Sun through even the smallest binoculars or telescope (see pp 192–201).

When compared with some other stars, the Sun is quite small, but compared with the Earth it is enormous (see pp 100–1) – having a diameter of 1,392,530 km (865,300 mls) and a mass about 330,000 times that of the Earth. The Sun's central regions are under great pressure and reach the tremendously high temperature of approximately 15 million° C (27 million° F).

In these conditions, nuclear reactions, in which hydrogen is converted into helium, are self-sustaining and form the source of the Sun's enormous output of energy. This energy is transported outward by radiation and convection until it reaches the visible surface of the Sun – the photosphere – where the temperature is about 5,800° C (10,470° F). From the photosphere, the energy is radiated out into space.

Although the temperatures throughout the Sun are so high that all matter is in gaseous form, its disc shows a distinct edge, known as the limb. Layers of the solar atmosphere lie outside this, but they are not normally visible, except during a total solar eclipse. At other times, the observer sees light from the lower, hotter layers at the centre of the disc and from higher, cooler layers at the limbs, which thus appear darker.

Far more conspicuous are the dark areas of the Sun's surface known as sunspots. These vary greatly in size, grouping, persistence and number. The smallest, known as pores, appear as tiny black dots. They may grow into larger spots that show a distinct structure, with a dark centre, the umbra, and a lighter outer region, the penumbra. The temperatures of the umbra

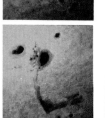

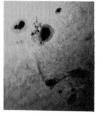

This sequence of six photographs of a small part of the surface of the Sun, *above*, was taken over a period of 40 minutes. Even over this relatively short time, changes in the dark filaments can be detected.

Prominences are easiest to see on the Sun's limb. When viewed in hydrogen light, they can also be seen against the full disc of the Sun. In this event they are known as filaments.

▷

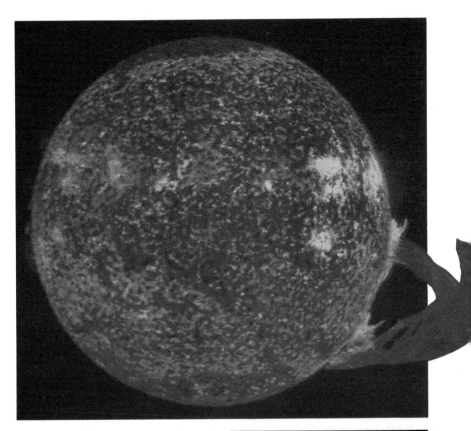

A gigantic solar prominence, *above*, was photographed from Skylab in December 1973. This is the image obtained at a single wavelength – that of singly ionized helium. The twisted structure visible in the loop follows the lines of magnetic force that arch out from the Sun's surface.

Ultraviolet observations, *right*, were computer-processed to give this image of the Sun with the same prominence shown, *above*. The colour coding emphasizes temperature differences. The coolest areas are light blue, followed by dark blue and purple. Active regions are red, and the hottest part of the flare is white.

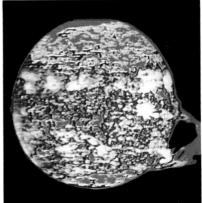

The Sun/2

and penumbra are about 3,800° C and 5,400° C (6,900° F and 9,750° F), respectively. The largest spots may cover areas that are far larger than the surface of the Earth.

With suitable, safe methods, you can make drawings of the full solar disc and of individual spots. Several drawings made over a few days will show the Sun's rotation. From Earth, this appears to take about 27·3 days at the solar equator and about 25·3 days relative to the stars. But since the Sun is not a solid body, different surface zones rotate at different speeds, although it may take several months of observation to determine this.

Much easier to follow is the way in which both individual sunspots and groups develop and change. Pairs of spots are quite common, as are the larger groupings, known as active areas. The latter are often dominated by a pair of large spots, accompanied by numerous smaller examples.

Records maintained over the years show that the number of sunspots varies considerably over an 11-year period. If the magnetic fields of pairs of sunspots are taken into account, it is found that the polarity reverses every 11 years. Strictly speaking, the true period of solar activity is, therefore, 22 years. A striking pattern emerges if the latitudes of sunspots are plotted over several cycles. The centres of activity gradually migrate toward the solar equator; as the last spots of the cycle reach low latitudes, the first spots of the next 11-year period begin to appear at high latitudes.

Faculae are bright patches of the solar surface which have a higher temperature than the surrounding area. Granulation describes the general overall mottling of the photosphere related to the convection cells in the outermost layer. Around the limb, glowing streamers and loops of gas, or prominences, can sometimes be observed. When they are in front of the

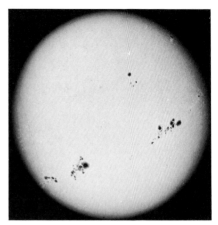

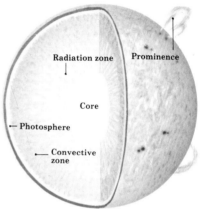

The interior structure of the Sun is simple. It consists of three main zones and the photosphere, *above*. Nuclear reactions produce energy only in the extremely small central core.

Radiation zone

Prominence

Core

Photosphere

Convective zone

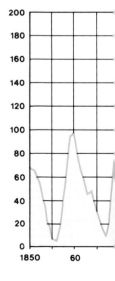

▷

Sunspots are revealed in a white-light photograph of the Sun, *left*. Two large groups lie on one side of the equator, which runs diagonally across the image.

A 'butterfly diagram' is a plot, such as the one, *below*, showing the positions of sunspots over a number of solar cycles. The general areas of activity can be seen to move slowly toward the Sun's equator. The plot shows also that high-altitude sunspots are rare.

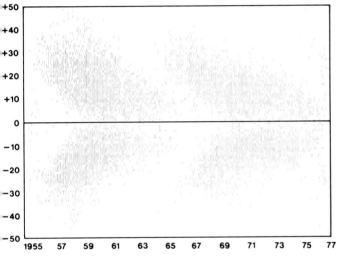

The **solar cycle** is shown, *below*, by the variation in the annual average number of sunspots. The 11-year cycle is obvious, but there are longer-term changes that are, as yet, only poorly understood.

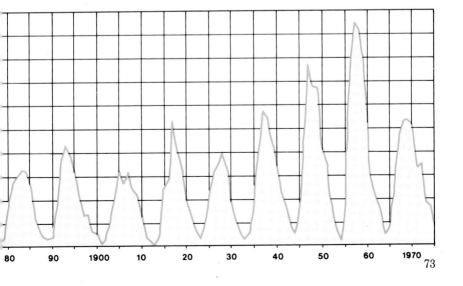

The Sun/3

disc they appear dark and are known as filaments. These are cooler than the photosphere itself and far colder than the layer in which they form; this is known as the corona – the outermost region of the solar atmosphere.

There is a general flow of charged particles (protons and electrons) out from the corona. This solar wind streams into interplanetary space, channelled by the lines of the Sun's strong magnetic field, which extends far beyond the orbits of the planets (see pp 102–3). Flares on the solar surface also eject additional streams of highly energetic particles from time to time, and these particles cause the Earth's aurorae (see pp 82–3).

The best conditions for studying the outer layers of the Sun's atmosphere occur when there is a total solar eclipse. Due to the inclination of the lunar orbit, an eclipse does not occur at every New Moon; however, between two and five solar eclipses can occur in any year (see pp 204–15). Unfortunately, even on these occasions, the disc of the Sun is rarely completely covered. A total eclipse takes place along only a narrow track of the Earth's surface; elsewhere, a partial eclipse is all that can be seen.

From Earth, the apparent sizes of the discs of the Sun and Moon are almost identical, but the Moon's elliptical orbit causes considerable variation. At its most distant, the Moon fails to cover the whole disc of the Sun and a ring of light remains; this is an annular eclipse. When the Moon is closest to Earth, a total solar eclipse (see pp 204–15) may last as long as 7 minutes 40 seconds. The normal duration is much shorter.

A total solar eclipse is spectacular. It begins slowly, since it takes about an hour from first contact for the Moon to cover the Sun, at second contact. In the early phases, the Sun remains so bright that only astronomers are aware that an eclipse is in

Moments from totality, this brilliant 'diamond ring' was photographed during the eclipse of 11 June 1983. The bright inner corona can be seen around the dark silhouette of the Moon, but the fainter outer regions remain invisible.

▷

The Sun/4

progress. Eventually, the change in the level of sunlight becomes noticeable. When only a thin portion of the solar disc remains uncovered, the landscape darkens and the birds and other animals react as if night were approaching. Finally, the shadow of the Moon can be seen rushing toward the observer.

Sometimes, just as the photosphere is covered, part of the chromosphere, with its characteristic pink colour, flashes into view. This layer is a region of rapid transition between the conditions in the photosphere and those in the corona. It is only a few thousand kilometres thick, and so is rapidly covered by the Moon.

At second contact – when the Moon covers the Sun – the corona becomes visible, shining with a distinctive pearly light. Its overall size and shape are closely linked with the stage of the solar cycle at which the eclipse is taking place. Around sunspot minimum, it tends to be reasonably symmetrical, and distinct solar plumes are usually visible. Near maximum, the corona is more extensive and may show long equatorial rays.

The short period during which the photosphere is covered is the only time it is safe to study any part of the Sun with optical equipment. Binoculars can then be used to pick out some of the detailed structure of the corona. These observations must cease the instant any part of the photosphere is uncovered (at third contact).

A brilliant, Diamond Ring effect often arises at either second or third contact, when just a tiny portion of the photosphere is visible. It is most notable when the contact point coincides with a lunar valley. Sometimes there are a number of irregularities in the lunar surface and a chain of brilliant points can be seen. These are known as Baily's Beads, after the astronomer Baily who first called attention to them in 1836.

The solar eclipse of 11 June 1983 is shown here in a sequence of pictures. The total eclipse was visible from Java, where these photographs were taken. The partial phase, *above,* was photographed using a solar filter.

The corona is seen at totality, *above.* Once the Moon has completely covered the Sun, photography without filters is possible. But the instant the Sun begins to reappear, filters must once more be used. Here an exposure of 1/15 sec shows the Sun's inner corona and the difference between its equatorial and polar regions.

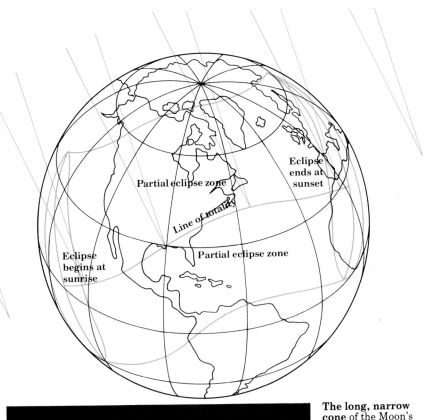

Partial eclipse zone

Eclipse ends at sunset

Line of totality

Eclipse begins at sunrise

Partial eclipse zone

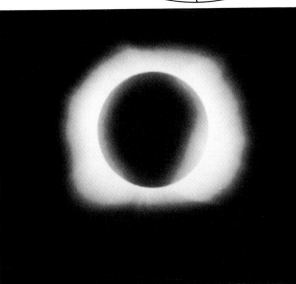

The long, narrow cone of the Moon's inner shadow, *above*, has an average diameter of about 160 km (100 mls) on the Earth's surface. A partial eclipse is visible from a much greater outer zone.

A longer exposure shows an enlarged corona, *left*. Because of the great range in brightness, the fainter outer regions (visible to the naked eye) were beyond the recording capabilities of the film used.

Mirages

Mirages are an interesting effect of atmospheric refraction (see pp 20–9). Although they are mostly associated with hot, desert conditions, mirages occur also when the temperature is extremely low. The most common mirage, which many people have seen, occurs when a heated road appears to be covered with water.

This type of mirage arises because the layers of air close to the Earth's surface have become warm and, therefore, less dense than higher layers. Rays of light close to the ground are bent sharply upward, and the 'water', which may show the 'reflections' of distant objects, is the image of the sky. This is known as an inferior mirage because the abnormal image appears below its true position.

Mirages that occur under cold conditions produce the opposite effect, and the images are visible above their positions. These superior mirages are caused when the lowest layers of air are colder than those above, so light rays are bent downward. This effect is best seen over the sea or an expanse of ice, which permit objects to be seen that are normally beyond the horizon. Sometimes, multiple images, one above the other, are visible.

Distant objects may also appear to be stretched vertically, an effect known as towering. Since height is one of the factors used by the brain to judge distance, objects affected in this way can seem either closer or more distant than they actually are. A particular form of towering is the Fata Morgana. This name derives from the sister of the legendary King Arthur, who, with her magical powers, was able to create castles out of thin air. The mirage appears as a vertical wall, building or high mountain range but, in reality, is an image of the sea, land or ice surface. The sharply bounded top is caused by an overlying inversion that cuts off light rays, preventing them from reaching the eye.

A 'lake of water' suddenly appears in a desert mirage, *above*. This photograph taken in an arid part of the Sahara near Tin Missao in Algeria, shows the classic form of this optical illusion. As the observer approached, the reflected image of the 'islands' and the sky shrank toward the vanishing line marked by the apparent water surface in the background. When the observer was extremely close, only the bare, dry hills could be seen.

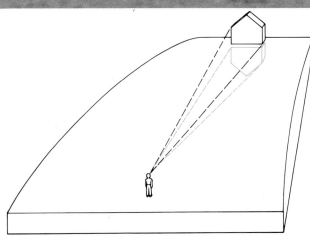

An inferior mirage always shows an inverted image of an object, *left*, formed as the rays of light pass across a strongly heated surface. The inverted image may sometimes be compressed vertically and then appears extremely small.

Nacreous and noctilucent clouds

Few clouds form above the troposphere. The vigorous convection currents in many cumulonimbus clouds often overshoot the tropopause and form a head of cloud above that level, which is where the anvils spread out. But true stratospheric clouds are rare because the air is so dry.

Just above the tropopause, thin clouds resembling ordinary cirrus and cirrostratus have been seen by observers in aircraft, but these have not been extensively studied. Even higher – between 20 and 30 km (12 and 19 mls) – strikingly iridescent clouds are observed on rare occasions. These nacreous, or mother-of-pearl, clouds are poorly understood but seem to be extremely high-altitude wave clouds (see pp 46–59). They are visible only around sunrise and sunset, when all lower clouds are in darkness.

At latitudes greater than about 45° North or South, during the months from late spring to early autumn, high clouds of another type may be seen on rare occasions. These are the noctilucent clouds, which are highly distinctive and can be seen shining above the horizon toward the poles in the hours around midnight. Their altitudes are about 80 to 85 km (50 to 53 mls) – three times as high as nacreous clouds. Their appearance is somewhat similar to cirrus, but they frequently show their own forms of rippled structure, which may be caused by fast airflow in high-altitude waves. The individual features often exhibit considerable motion during the course of a display.

Noctilucent clouds pose many problems and are something of a mystery. Doubt still remains about the mechanisms by which these clouds are formed, since the air at such great altitudes is dry and generally thought to lack freezing nuclei (see pp 40–3). One suggestion is that the nuclei are particles of meteoric dust. It appears improbable for freezing nuclei to be transported to such extreme heights.

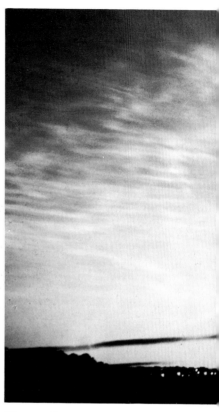

The noctilucent clouds in this display, *above,* were photographed in Scotland at midnight, looking north. The regular pattern on the left

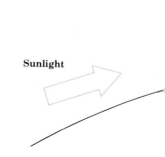

Sunlight

and the more random structure over the rest of the display are both typical of this form of cloud. The top of the twilight arch is distinctly visible. **Close to the Equator** the Sun does not illuminate any visible portion of the atmosphere around midnight, as can be seen, *below*.

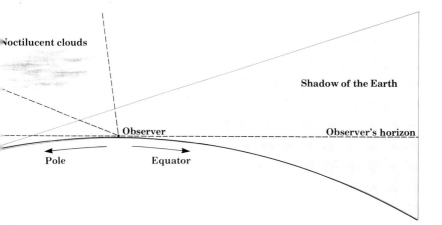

Noctilucent clouds

Shadow of the Earth

Observer

Observer's horizon

Pole   Equator

Aurorae

A major display of the aurora borealis, the northern lights, or its southern counterpart, the aurora australis, is an unforgettable sight. It may begin quietly, perhaps an hour or so before midnight, with the appearance of a faint glow on the horizon or of small, structureless patches of light (P).

These may develop into a smooth arc of light, stretching across the northern sky from east to west in the northern hemisphere. Such an arc (A) always has a fairly sharply defined lower edge and a much more diffuse upper border. If the arc is smooth, it is described as being homogeneous (HA). Individual rays (R) may, however, become visible, or the whole display may develop this type of structure and become a rayed arc (RA). Sometimes, curtainlike folds develop, described as a band (B), which may be homogeneous (HB) or rayed (RB). In a strong display multiple bands may be present, and if the aurora occurs overhead it may present a fanlike structure, known as a corona (C).

Some displays are quiet, showing only a single form all night. Others may change rapidly, undulating and surging across the sky. Faint displays appear colourless or pale green. The strongest and most striking aurorae may be bright red or show distinct yellow and bluish tinges.

Aurorae are most frequently seen in two zones, known as auroral ovals, lying around the magnetic poles. It is here that the Earth's magnetic field allows charged particles, which originate in the solar wind (see pp 70–7), to enter the upper atmosphere. There the particles excite molecules of gas to higher energy states, and the gas radiates the additional energy, producing the visual display. Auroral activity is closely linked to that of the Sun and shows a similar 11-year cycle. The strongest displays occur after solar flares have injected numerous energetic particles into the solar wind.

A brilliant auroral corona, *above,* was photographed when almost overhead. The observer is looking directly along the lines of the Earth's magnetic field, so the rays appear to converge on a single area of the sky. Such a corona is often the most spectacular part of a display. It gradually dies away toward dawn.

Observations of the aurorae are easiest in North America because the auroral oval reaches lower latitudes than it does over Europe. The fine display of auroral bands, *right,* was photographed in Alaska.

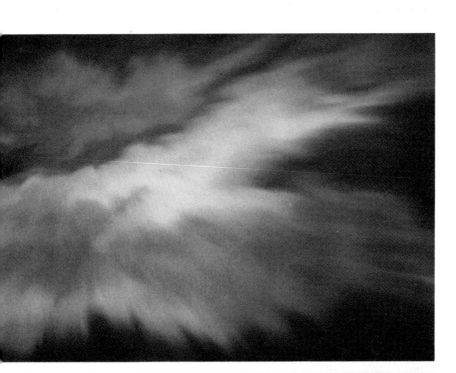

The sky from above/1

A flight in an aircraft is an ideal opportunity for making meteorological observations of effects that are rarely seen from the ground. Clouds are, of course, the most obvious objects visible. Their relative height and thickness are frequently difficult to estimate from the ground, especially when several layers are present.

The relationships between the various layers and cloud types usually become clear as the aircraft climbs to its cruising height. Even with scattered cumulus, the uniform height of the condensation level is apparent. Similarly, once the aircraft is above a low cloud layer, it is possible to see the way an inversion limits the clouds' upward growth. Certain strong convection cells may be breaking through the inversion and building individual cumulonimbus heads that are reaching much higher levels. It is difficult to tell from below if this is happening when, for example, there is extensive stratocumulus cover.

Turbulence at take-off and landing is a result of hot-air thermals rising from the ground. A large number of cumulus and cumulonimbus clouds is a sure sign that convection is active.

The cruising altitudes of most aircraft are above the level of most clouds, except for the highest cumulonimbus. Aircraft may sometimes fly through the tops of these clouds, but wherever possible, pilots try to avoid the severe conditions in the lower regions of cumulonimbus.

If your flight takes you across frontal systems, you should be able to identify the changes in the types of clouds. Cloud variation is most marked in low-pressure systems. In summer, when the land is hotter than the oceans, numerous cumulus and cumulonimbus build up over the land while the sky above the sea is more or less clear.

In the colder months, when conditions are reversed, it is often the air

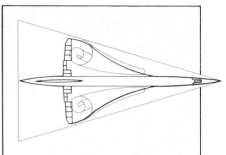

Aircraft vortices
The leading edges of the wings of all aircraft produce low-pressure areas, where condensation may occur. The pair of vortices may be extremely dramatic with swept-back wings, such as those on the supersonic Concorde. The curling sheet of condensation is particularly noticeable on take-off.

An aerial view,
right, reveals clouds brought in by the monsoon to the foothills of the Himalayas. In the background, sheets of cloud can be seen over the lower plains. Flights across major mountain barriers often reveal startling differences in the cloud cover on the two sides of the same range.

▷

The sky from above/2

over the warm water that is unstable and produces shower clouds. This also happens frequently when a cold front draws cold polar air over a large body of water, such as a lake or the sea.

Looking down, it is most noticeable that sea or lake fog often covers only the water and neighbouring shores, while farther inland the weather is fine and clear. Inland fog, on the other hand, frequently hides low-lying areas but leaves the higher hills and mountains in sunshine. The higher clouds, produced by moist air being lifted over mountain areas, can often be seen from a great distance, as can wave clouds that form downwind.

Both haloes and rainbows can take on different forms when seen from the air. The additional elevation makes it possible to see a complete circular rainbow, for example. As well as the usual halo phenomena, you may be able to see a 38° halo in the direction away from the Sun, as well as a brilliant subsun that forms at the anti-solar point. A number of lower arcs, normally invisible from the ground since the ice crystals have to be below the observer, can also be seen.

On the side of the plane away from the Sun, it is possible to see a glory if the flight is above a bank of clouds. When the aircraft is close to cloud level, the glory's shadow mimics that of the plane. Farther away from the clouds, the shadow shrinks, becoming first a silhouette of the aircraft and then a tiny circular patch. At an even greater distance, the shadow disappears and may be replaced by a bright spot. This resembles *heiligenschein*, which can sometimes be seen on dew-covered grass surrounding the shadow of an observer's head, and results from light being reflected back directly toward the Sun.

Aircraft create their own clouds in the form of condensation trails, or contrails. Two significant processes are involved in this. Although the air

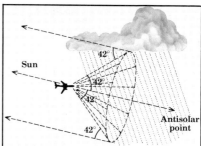

Rainbows from the air
Travellers in aircraft can often see a rainbow as a perfect circle with a diameter of 42°, centred on the anti-solar point. As when viewed from the ground, the bow's exact position depends on the elevation of the Sun. On occasions, only the lower portion of the circle is visible.

▷

Layer cloud shrouds Scottish mountains, *above*, in a photograph taken from a fairly low altitude. In this region, airflow around a stationary high-pressure region to the west often brings much low cloud, but the tops of the highest peaks may be in clear air above an inversion.

Cumulus and stratus clouds can be seen, *right*, over the sea. The bright colour fringes are produced by the polarizing effects of the plastic material in the aircraft window and are made visible by a polarizing filter on the camera lens.

The sky from above/3

at any particular level may well be saturated with water vapour, it may contain so few nuclei that no condensation takes place until vast numbers are introduced by the exhaust from the jet engines. Additionally, the exhaust gases themselves contain a considerable quantity of water vapour, which is one of the products of combustion. Since contrails are not produced if the air layer is exceptionally dry, they sometimes end abruptly when the plane enters dry air. Such breaks can occasionally be seen in the shadow of the aircraft contrail cast on the clouds below.

Any aircraft generates various low-pressure regions in the airflow around it, where localized plumes of condensation are produced if the humidity is high. With a propellor-driven aircraft, low pressure occurs at the tip of every propellor blade, with the result that as the blades sweep through the air, they form heliacal trails of condensation.

All aircraft may trail condensation plumes from their wing tips. Here, the condensation takes place in the low-pressure core of the vortex generated by each wing. The two, normally invisible, counter-rotating vortices trail behind the aircraft, just like the wake of a ship. The exhaust from the engines is swept into this rotation, which explains why contrails always consist of a pair of rolls of condensation, regardless of the number of engines. The type of wing-tip vortices produced by large aircraft are very powerful, and may pose a serious hazard for smaller craft.

If the aircraft is flying above a layer of thin cloud, these vortices may force warmer air down from the higher level. This can be sufficient to disperse the cloud beneath the path of the aircraft, so forming a distrail. On other occasions, the aircraft may initiate freezing in the cloud, producing a clear channel when the ice crystals fall out into the lower air.

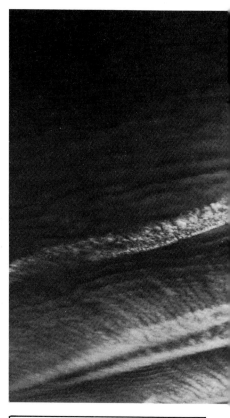

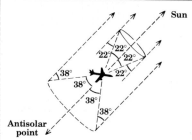

Halo phenomena
These may be seen below an aircraft, as well as the normal effects visible from the ground. The 38° halo and a subsun – a bright elliptical spot at the antisolar point – are the most common.

Contrails often
indicate the
passage of a jet
aircraft, *above*. The
persistence of the
trails probably
indicates the
approach of a
depression.

**All mountain
shadows** appear
triangular,
whatever the
mountain profile,
as in the example,
right, of Mauna
Kea, Hawaii. The
shadow edges
appear to converge
only because of
perspective.

Weather lore

Every country has a vast store of weather lore enshrined in traditional rhymes and sayings. Even within one country, many are contradictory and others quite unreliable. Some are based on old, erroneous notions, such as the idea that the crescent Moon 'holds water in its lap' that will spill out when the horns of the crescent are tilted toward the ground. Much lore has survived because only the occasions when it appears to come true are remembered—the greater number of times when it fails are forgotten.

Some old beliefs do, however, have a sound basis in fact. The idea, for example, that a halo is a sign of impending wind and rain dates back to observations made by the ancient Chinese. Haloes occur in cirrostratus, which is usually a sign of an approaching depression. The forecast of high winds following widespread mares' tails (cirrus) is similarly correct.

The belief that thunderstorms end a spell of fine weather is certainly not true on many occasions. Thunderstorms occur only in an unstable air mass, but after they have taken place stability may be increased, so the fine weather is restored or even improved.

In the case of lightning it could be dangerous to base your behaviour on some well-known sayings, especially those that claim that one type of tree is less likely to be struck than another, or that lightning never strikes twice in the same place. Whatever the species, any isolated tree, or building, is a dangerous sheltering place because, being the highest point in the vicinity, it is the most likely to be struck, often more than once.

Longer-term predictions are rarely sound. Many, such as that a plentiful crop of wild berries foretells a harsh winter, say more about conditions earlier in the year when the plants were in flower. The weather is so complex that accurate predictions for even a few hours may be difficult at times.

'A red sky at night' does indeed frequently indicate that the next day will be fine. It shows that the sky to the west is fairly clear of cloud. At sunrise, the same tints suggest that high cloud is spreading from the west and that a rainbelt will follow during the day. Yellow coloration late in the afternoon often means that the atmosphere is moist and that rain will move in during the night.

Building a weather station/1

For an amateur skywatcher, keeping a regular record of weather observations can become an absorbing pastime. Initially it is not necessary to have a lot of expensive equipment, although you may wish to add some instruments at a later date.

Even without instruments, general descriptions of the weather and observed phenomena can be informative. The type and amount of cloud present, wind direction, whether there has been rain or snow and similar information can be noted. Some observers also keep details of the dates on which particular plants come into bud or leaf, blossom, and drop their leaves. Although only one of the factors, such records can provide an insight into the connection between weather patterns and the entire natural world.

The most important instruments to acquire are a barometer and suitable thermometers. A barograph, however, will relieve you of the task of having to read a barometer at fixed times and will also provide an informative, continuous record, showing many minor variations that would otherwise go unrecorded.

For temperature measurement, it is probably best to start with an ordinary maximum and minimum thermometer. This will show the temperature at any time and will automatically record the highest and lowest temperatures that have been reached since the thermometer was last reset. You may want to add a simple pair of matched wet- and dry-bulb thermometers, which, with the use of standard tables, will enable the humidity to be determined.

There is no need to buy expensive, fully tested and certified instruments of the type required by professional and voluntary observers who operate recognized meteorological stations; moderately priced versions will prove quite satisfactory for most general-interest purposes. Nevertheless, buy

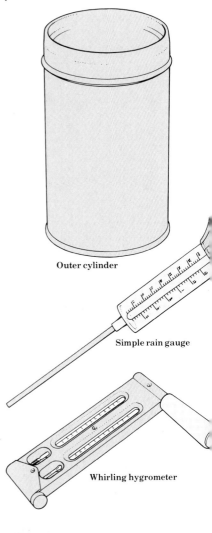

Outer cylinder

Simple rain gauge

Whirling hygrometer

An amateur weather station can be started with a few simple instruments. In the maximum and minimum thermometer, an index moves in each branch of the tube to show the highest and lowest

▷

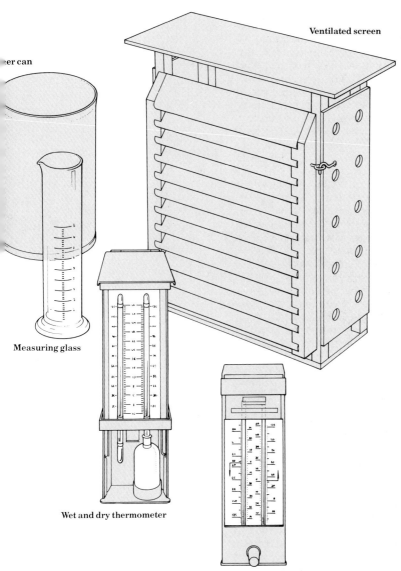

er can

Ventilated screen

Measuring glass

Wet and dry thermometer

Maximum and minimum thermometer

temperatures reached. Wet and dry thermometers can be used to calculate air humidity, as can the whirling hygrometer.

A rain gauge includes an outer cylinder with a collecting funnel, an inner can and a measuring glass. A simple version incorporates a graduated scale. Thermometers should be mounted in a ventilated screen, which allows the free flow of air but excludes direct sunlight.

93

Building a weather station/2

the most accurate equipment you feel you can afford.

Since thermometers must record actual air temperatures, they should be mounted in a specially made screen that protects them from direct sunshine but still allows air to circulate freely. Simple screens are available as kits for home assembly. Alternatively, it is possible to make your own screen, which should be painted white to prevent it from absorbing too much heat from the Sun and should be firmly mounted on a strong stand so that it is not damaged by high winds. The door must be on the north side (in the northern hemisphere) so that sunshine does not affect the thermometers while readings are being taken.

Measurement of the wind poses many problems. Any obstructions, such as buildings or large trees, can affect the speed and gustiness of the wind, as well as the direction as shown by a simple wind vane. Apart from being turbulent, the lowest layer of the air is also slowed by friction with the ground. For these reasons, professional anemometers are placed on masts 10 m (33 ft) high. For most amateurs, anemometers are too expensive and impracticable to install. For these reasons, it is best to determine the direction of the wind by watching the movement of the lower clouds, and to learn to estimate its strength by using the Beaufort scale (see p 35).

Measurement of rainfall can also be difficult. Rain gauges must be sited clear of obstacles so that raindrops do not splash into, or out of, the collecting funnel. The amount of water collected when there has been torrential rain or windy conditions should be treated with some reservations. And when snow falls, the quantity collected by the rain gauge should be melted gently to give the amount of rainfall equivalent. Snowfall is so greatly affected by any wind gusts, however,

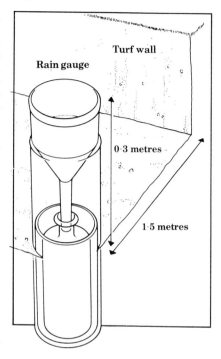

Rain gauge

Turf wall

0·3 metres

1·5 metres

A rain gauge must be sited with care. Its surrounds should be clear of obstructions. Any that do exist must be a distance of at least twice their height away from the gauge.

The glass sphere in a sunshine recorder concentrates the light of the Sun into a single point, burning a trace on a card held in the curved frame. The instrument must be adjusted so that its angle corresponds to the latitude of the observing station.

▷

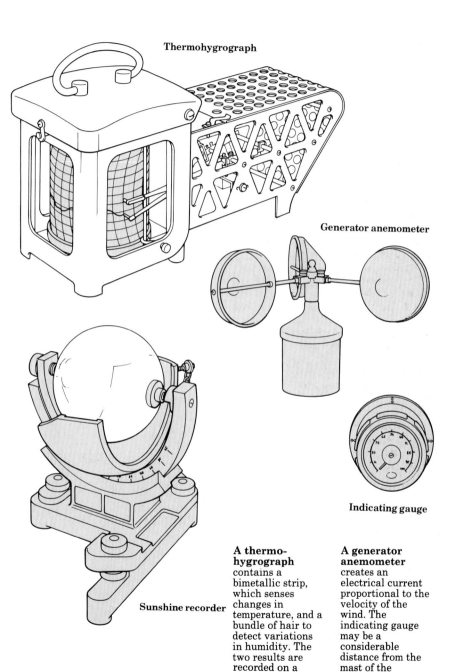

Thermohygrograph

Generator anemometer

Sunshine recorder

Indicating gauge

A thermo-hygrograph contains a bimetallic strip, which senses changes in temperature, and a bundle of hair to detect variations in humidity. The two results are recorded on a single chart.

A generator anemometer creates an electrical current proportional to the velocity of the wind. The indicating gauge may be a considerable distance from the mast of the anemometer itself.

Building a weather station/3

that the results must be regarded as approximate.

Make readings of instruments regularly, but choose the times that will best suit your other activities. Ideally, observations should be made on the hour. Official stations do this just before the hour, ending with the pressure reading, which is made as close as possible to the nominal reporting time. Greenwich Mean Time (GMT) is used as the international standard; it exactly corresponds to Universal Time (UT) used by astronomers (see pp 202–3). Always keep to GMT, so if you make observations at 8.00 am and 8.00 pm (8.00 hrs and 20.00 hrs GMT) during the winter, make them at 9.00 am and 9.00 pm clock time when an additional hour has been added for Summer Time.

Records should be entered in a permanent log book. Apart from actual instrument readings, weather and cloud observations may either be described or entered, using the standard symbols (see pp 44–5). Even if you can make readings only once or twice a day, try to look at the sky from time to time. This will help to make your weather records more complete. Make sure that any unusual events and observations, such as haloes, mirages, aurorae and so on, are recorded in full, with the exact time, position in the sky and any other relevant information.

For a fuller and more complete understanding of the often erratic nature of the weather, and the way it changes from season to season, it can be useful to make both monthly and yearly summaries of your observations. These can record information such as extreme and mean temperatures and pressures, the number of days with rainfall or winds of a particular force or direction. Much of this information can then be drawn up in the form of simple graphs and charts, which are helpful when you wish to compare current and prior observations.

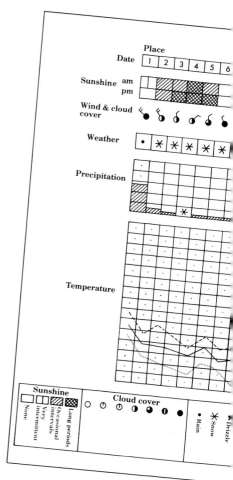

A graphical record summarizing the weather over a period of a month has been recorded. If circumstances permit, observations should be made and recorded several times a day. It may not be practicable to show all the information on a summary such as this.

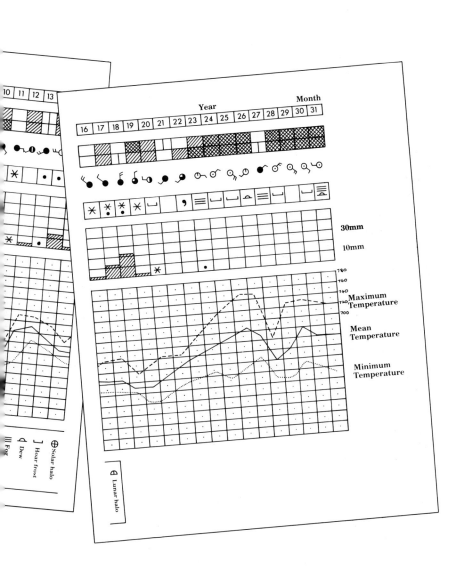

Year   Month

| 16 | 17 | 18 | 19 | 20 | 21 | 22 | 23 | 24 | 25 | 26 | 27 | 28 | 29 | 30 | 31 |

30mm

10mm

Maximum
Temperature

Mean
Temperature

Minimum
Temperature

⊕ Solar halo
] Hoar frost
◖ Dew
‖‖ Fog

◖ Lunar halo

97

N I G H T   S K Y

Approaching the sky

If you have the opportunity, lie down on the ground in a large park or out in the country and look up at the night sky. It appears as if the Earth is floating in a vast starry universe that stretches out in every direction.

In fact, even under ideal conditions, you can never see more than about 2,500 stars at any one time. There are many more stars than this, but they are too dim to see without binoculars or a telescope (see pp 192–201).

Continue watching the sky for a few hours and you will notice that the stars are not still. They appear to move slowly across the sky from east to west; this movement is 15° every hour, which means that the whole sky seems to rotate once every 24 hours. All the stars take part in this daily (or diurnal) motion, retaining their same positions relative to one another as if fixed on the inside of the spinning sphere, with the Earth suspended at the centre.

Detailed observations show that there is no spinning celestial sphere and that the stars are spread out in space; it is the Earth that is rotating. We are moving observers who see the skies seeming to rotate only because of our shifting position in space.

The planets can be divided into two groups. One group consists of the solid bodies which have either no atmosphere or one not very thick. The other contains the gas giants with extensive atmospheres. These include Jupiter, Saturn, Uranus and Neptune. The diameters, *right*, include the core and atmosphere.

Photographed from the Apollo II orbiter, this picture, *opposite*, shows the Earth rising above the surface of the Moon.

The distances of the planets from the Sun, *below*, are expressed as an average because their orbits are not exact circles. The massive size of the Sun relative to the planets is evident.

Planetary diameters

	km	mls
Mercury	4,878	3,031
Venus	12,104	7,521
Earth	12,756	7,927
Mars	6,796	4,223
Jupiter	142,796	88,733
Saturn	120,660	74,978
Uranus	50,800	31,567
Neptune	48,600	30,200
Pluto	3,000	1,864*

*Value slightly doubtful

Mean distance from the Sun   (km / million)

57·9	108·2	149·6	227·8	778·3

Mercury   Venus   Earth   Mars   Jupiter

This rotation of the Earth is not unique. All bodies in the universe – stars and planets – are rotating. The stars appear to be fixed in space only because they are so far away that their individual motions are too small to detect without the most refined techniques. But moving they certainly are, orbiting around the centre of our own giant star system – our Galaxy.

The fact that the other planets orbit the Sun as the Earth does means that from our moving vantage point they appear to shift their positions against the background of the stars. This, indeed, is why they are called planets, a name derived from the Greek word for wanderers.

In common with the other planets, but not the Sun and other stars, the Earth is a cold body. It does not shine in space as a star would. Certainly it is visible from out in space, as photographs from space probes and the photograph of Earthrise taken from the Moon have shown, but this is only because the Earth reflects light from the Sun. And what is true for the Earth is true also for the other planets that orbit the Sun and together form most of what has come to be known as the Solar System.

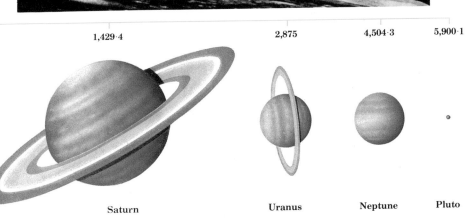

1,429·4   2,875   4,504·3   5,900·1

Saturn   Uranus   Neptune   Pluto

The Solar System

The Solar System is like a small oasis in space. At its centre is the Sun, whose gravitational pull keeps the planets in their orbits. These orbits, except for that of Pluto, lie in nearly the same plane; so if you were to make a model of the Solar System, all the planetary orbits would be contained within the thickness of a disc like a phonograph record.

All the planets orbit the Sun in the same direction; looked at from a position above the Solar System, they move counter-clockwise. The speed at which each planet travels depends on its distance from the Sun, those closer moving faster than those farther out. Thus Mercury, the planet closest to the Sun, moves at a speed of 47.87 km/sec (29.75 mls/sec), whereas Pluto, the outermost of the planets, travels at only 4.74 km/sec (2.95 mls/sec). So while Mercury takes only 88 Earth days to complete one orbit of the Sun, Pluto takes 248.5 Earth years – more than a thousand times longer.

Another feature common to all the planets is that they rotate about their axes as they orbit the Sun. One complete axial rotation is a day, but this period is highly variable. On Venus, for example, axial rotation takes

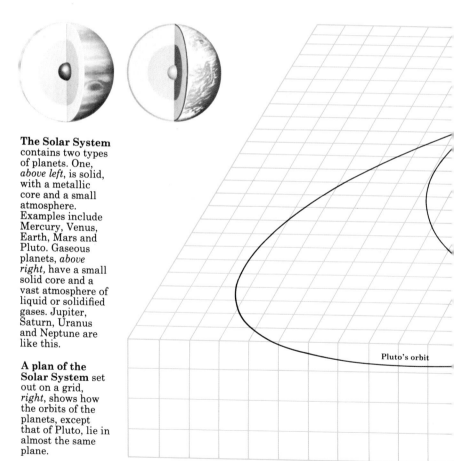

The Solar System contains two types of planets. One, *above left*, is solid, with a metallic core and a small atmosphere. Examples include Mercury, Venus, Earth, Mars and Pluto. Gaseous planets, *above right*, have a small solid core and a vast atmosphere of liquid or solidified gases. Jupiter, Saturn, Uranus and Neptune are like this.

A plan of the Solar System set out on a grid, *right*, shows how the orbits of the planets, except that of Pluto, lie in almost the same plane.

Pluto's orbit

approximately 243 Earth days, while on Jupiter a complete rotation takes a little less than ten hours. Also varying is the degree of axial inclination of the planets. For example, the Earth's polar axis is inclined at an angle of 23° 27′ to the vertical (the vertical being measured with respect to the plane of the Earth's orbit). For Venus, however, the axis is inclined at 178°, while the axial inclination of Jupiter is a mere 3° 05′.

As well as these larger, named planets, none of which has a diameter smaller than one-quarter of the Earth's, the Solar System contains a host of other bodies. Concentrated mainly between Mars and Jupiter, there are tens of thousands of asteroids that orbit the Sun, the largest of which is Ceres, with a diameter of 974 km (605 mls).

Also in the Solar System are meteors and comets. Comets have elongated elliptical orbits which lie at quite wide angles to the plane of the Solar System. If the size of the Solar System is taken as being the distance these comets travel from the Sun before starting to return, then the Solar System extends some two light-years out into space.

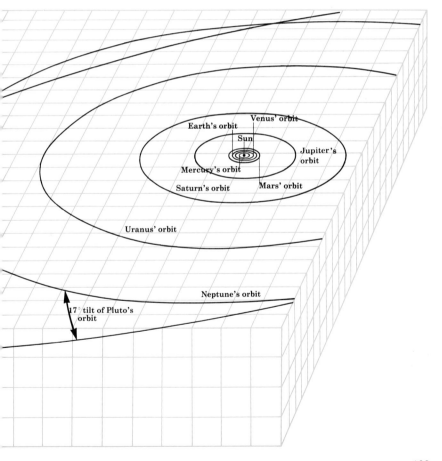

Mercury and Venus

Mercury and Venus have inferior orbits (orbits that keep them closer to the Sun than to the Earth). Because they are closer to the Sun than we are, these planets always appear in the sky comparatively near to the Sun, and can only be seen a little before sunrise or a little after sunset. For this reason, both Mercury and Venus are sometimes called the Morning Star or the Evening Star.

When seen through a telescope, both planets show phases similar to those of the Moon. When close to Earth, they present a crescent phase; they appear as discs only when on the far sides of their orbits. Thus it is difficult to see detail on either planet.

Both planets are bright and can sometimes be seen in daylight if you know just where to look. But never try if they are near the Sun. Mercury can sometimes almost equal Sirius, the brightest star in the night sky and, on occasions, Venus can appear 10 times brighter than Mercury.

Another observational problem is that, appearing as they do, close to the horizon, the light from both planets must travel through a thick layer of Earth's atmosphere. This could account for the fact that Mercury

The orbits of Mercury and Venus, *right*, lie closer to the Sun than that of the Earth. These planets thus show phases. The discs of both planets cannot be seen in full at any one time because they then lie behind the Sun. When the major part of the disc is seen, the planet lies at almost its farthest point from Earth. The two planets always appear close to the Sun, *middle right*, because their orbits are closer to the Sun than that of Earth.

Mercury, *far right*, was photographed in space by Mariner 10. The phases of Venus, *right*, seen through a telescope, show how difficult it is for an observer on Earth to detect any surface detail.

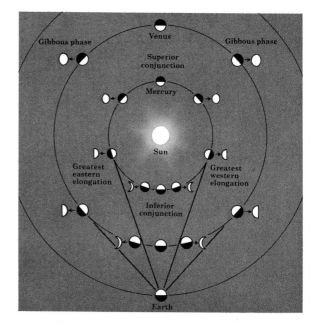

Mercury Venus Earth

twinkles and is sometimes observed to have a slightly pinkish colour.

When Mercury is in transit, that is, directly in line with the Earth and Sun, it can be observed as a black spot against the face of the Sun. Mercury is too small to see without a telescope and, because it has no atmosphere, its outline appears crisp with no distortion. The next transit of Mercury will be on 12 November 1986 and then not again until 5 November 1993. But never view this event directly through a telescope or you could damage your eyes (see pp 182–3).

Transits of Venus are less frequent than those of Mercury, but during transit Venus is clearly visible to the naked eye against the backdrop of the Sun. The dense atmosphere of Venus does, however, cause some distortion to the appearance of the disc. The next two transits of Venus are not, unfortunately, until 2004 and 2012.

Taking the distance from Earth to Sun as one unit (known as the astronomical unit or simply a.u.) then the average distance between the Sun and Mercury is 0.387 a.u., and between the Sun and Venus, 0.723 a.u. These distances vary because neither planet moves in a precisely circular orbit.

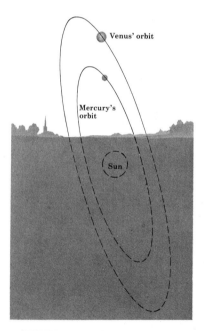

Mars

Mars is the first of the superior planets: those with orbits farther away from the Sun than that of the Earth. It can thus be seen at any time during the night, as long as it is above the horizon. Mars is best observed at opposition, when it is directly opposite the Sun as seen from Earth, which occurs every 26 months; it can then be as bright as Sirius.

Not all oppositions, however, are equally good. The best is when Mars is at or near perihelion: its closest approach to the Sun. Perihelic oppositions occur about every 16 years, and, for European and American observers,

when Mars is in opposition and placed south of the celestial equator, it is less well situated for observing.

Mars is not large, being intermediate in size between Mercury and Venus, and it is not an easy planet on which to observe detail, although better in this respect than either Mercury or Venus. Its most distinctive characteristic is its obvious red colour. The problems of detecting detail arise solely from its size and distance from Earth – never does its disc appear larger than about 24 arc seconds (1 arc second = 1/3,600 of a degree). In a pair of binoculars Mars will always appear

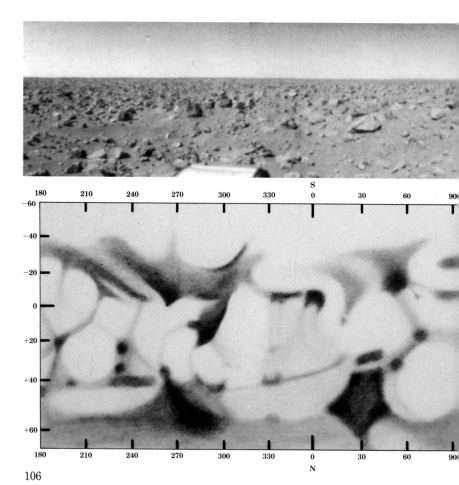

extremely small, but in a telescope of 15-cm (6-in) aperture, such as many amateurs possess, it does display a little detail – Syrtis Major being the easiest landmark to pick out.

Even in large telescopes the surface features of Mars can never be observed with great precision. That Mars possesses mountains and craters, rifts and valleys, was discovered only after spacecraft such as Mariner 9 went into orbit around the planet in 1971–2. All the same, a telescope will show you that Mars has white polar caps, that its surface is reddish in colour and that it has also some blue-grey areas.

The idea that Mars was crossed by a vast system of artificial canals has proved to have been based on an optical illusion due to the difficulties of observing surface detail from Earth.

Amateur astronomers can still do much useful work observing Mars because, although they cannot detect a great amount of detail, they can follow the large dust storms that sweep over the planet. They can thus monitor atmospheric changes when there is no spacecraft near by to observe them. Mars has two tiny moons, Phobos and Deimos, but neither can be seen with amateur equipment.

The planet Mars was photographed, *left*, by Viking 2, which landed on the Martian surface in September 1976. This area is comparatively flat, but Mars has some mountainous regions.

A drawing, *above*, shows Mars and the orbits of its two satellites.

A map of Mars, *left*, made by an amateur astronomer, is very different from a map made using photographs taken from spacecraft, but it does delineate the features that can be seen from Earth. From these features an observer can detect changes as the planetary seasons came around. Amateur observations contribute greatly to studies of any planet.

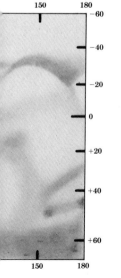

Mars, seen through a telescope, was the basis for the amateur drawing, *below*, that shows the dark area Syrtis Major.

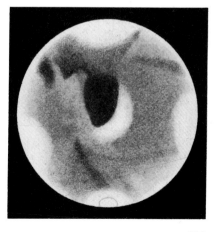

Jupiter

Jupiter was the chief of the Roman gods and his name was given to the planet because it is by far the brightest object in the Solar System except for Venus, which can sometimes outshine it. But Jupiter, the innermost of the gas giants, can certainly be by far the brightest of the planets to be seen in the night sky. It can be up to three times brighter than Sirius, and its steady light makes it a magnificent sight in the night sky.

Using any telescope or good binoculars, you can see Jupiter's disc as well as its four largest satellites – Io, Europa, Ganymede and Callisto. At opposition, which occurs approximately every 13 months, Jupiter presents a disc some 10 times larger than Mars. This is because of Jupiter's immense size; it is the biggest planet in the Solar System, more than 11 times larger than the Earth, and is greater in mass than all the other planets together.

In a telescope, Jupiter displays a series of coloured bands across its disc, parallel to its equator; these are currents in its thick, cool atmosphere. The more stable of these currents have been given names (see below). Often, too, there are easily discernible spots

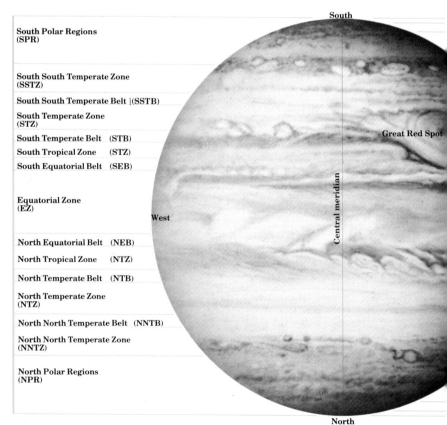

South

South Polar Regions (SPR)

South South Temperate Zone (SSTZ)

South South Temperate Belt (SSTB)

South Temperate Zone (STZ)

South Temperate Belt (STB)

South Tropical Zone (STZ)

South Equatorial Belt (SEB)

Equatorial Zone (EZ)

West

North Equatorial Belt (NEB)

North Tropical Zone (NTZ)

North Temperate Belt (NTB)

North Temperate Zone (NTZ)

North North Temperate Belt (NNTB)

North North Temperate Zone (NNTZ)

North Polar Regions (NPR)

Great Red Spot

Central meridian

North

Earth

and other changes in the atmosphere. When you look at the whole disc, the planet appears slightly 'squashed' because its relatively fast rotation causes the extensive atmosphere of the equatorial region to bulge. Do not be disappointed if initially you see little detail; allow time for your eyes to adjust, and more and more detail will become evident.

One of the most astounding sights on Jupiter is the Great Red Spot; this is at least 300 years old and sometimes seems to disappear. It is probably an atmospheric whirlpool and is often noticeably red in colour. Like the other features in the Jovian atmosphere, it moves across the disc and is not difficult to follow even through a low-power instrument.

You can also observe the equatorial regions rotating faster than the polar regions; at the equator, rotation takes 9 hours, 50 minutes, 24 seconds, whereas the north polar region takes some 5 minutes, 18 seconds longer. When looked at through a telescope, the north polar region is the lower part of the disc, but it is the upper part when looked at through binoculars. This is because an astronomical telescope inverts the image.

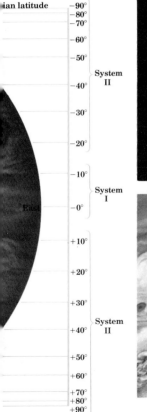

ian latitude

-90°
-80°
-70°
-60°
-50°

System II

-40°
-30°
-20°
-10°

System I

-0°
+10°
+20°
+30°

System II

+40°
+50°
+60°
+70°
+80°
+90°

Jupiter's thick atmosphere has been charted by observers, *far left.* The zones are used as a referral base for observed changes. Jupiter and the four bright Galilean satellites are shown in the photograph, *above.* A photograph taken by Voyager 2 from 5.92 million km (3.72 million mls) away, *left,* shows the semi-permanent Red Spot.

Saturn

None of the gas giants is dense, but Saturn is the lightest of them all, having only about half the density of Jupiter. If there were an ocean big enough, Saturn would float in it. It was the most distant planet known before the days of the telescope, but when seen through one, Saturn with its rings is probably the most magnificent sight in the night sky. It is, however, nowhere near as bright as Jupiter. Even at opposition, which occurs at intervals of just over one year, it is never more than about one-fifth as bright.

Saturn is only a little smaller than Jupiter (its diameter is 20 per cent less), but it appears much smaller in a telescope because it is almost twice as far away. For detailed observations of Saturn's atmospheric belts, which are not so prominent as Jupiter's, you will need a telescope with an aperture of 20 cm (8 in) or even 25 cm (10 in). The system of rings circling the planet in the equatorial plane is, however, readily seen in some detail in a smaller instrument – a 15-cm (6 in) telescope should be sufficient for this.

Saturn offers important work for amateur astronomers. The planet is the second of the Solar System's four

An observer's drawing of Saturn, as seen through a 25-cm (10-in) telescope, clearly shows the planet's full disc and rings. The photographic close-up of the ring system, *left,* was taken by the Voyager 1 spacecraft on 13 November 1980 at a distance of 1.5 million km (930,000 mls) from the planet.

Earth's view of Saturn's rings changes constantly, *opposite*. This occurs because the rings are tilted with respect to the plane of the Earth's orbit. As a result, the underside of the rings can be seen at some times; at others, the top is visible. Because they are so thin, when the rings appear edge-on they seem completely to disappear.

Saturn   Earth

gas giants and, since it is largely gaseous, different parts of the globe rotate at different speeds. Much more detailed and constant observation of Saturn is required to help refine what is known of these rotation periods. You will find that a light blue filter used over the telescope eyepiece will help sharpen image definition.

As the relative positions of Earth and Saturn change, so the angle at which the rings are seen alters. Every 13 years 9 months and 15 years 9 months alternately, the rings are edge-on and disappear from view. The last time this occurred was in March 1980; it will happen again in September 1995. Midway between these times, in June 1988, the rings will appear wide open.

Observation through a telescope shows that the rings have divisions – most noticeable is the Cassini division, named after its seventeenth-century discoverer. This divides the outermost bright ring (ring A) from the next bright one (ring B). Ring A is itself divided by Encke's division.

Although Saturn has 17 satellites, only Tethys, Diona, Rhea, Titan and Iapetus can easily be seen. They are best seen when the rings are edge-on.

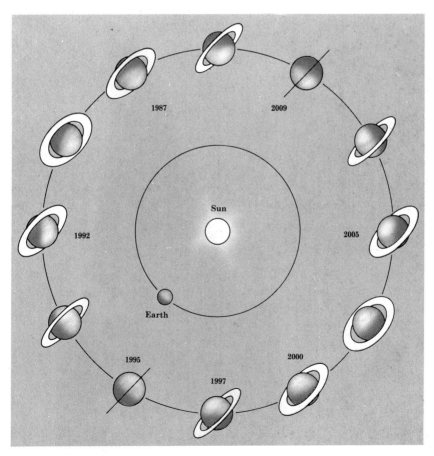

Uranus, Neptune and Pluto

The three planets of the outer Solar System were not known in ancient times. Uranus – the first planet to be discovered in modern historical times – was observed in 1781 by William Herschel. Irregularities in the orbit of Uranus led to Neptune's discovery in 1846, though its existence had been predicted the year before by the Englishman John Couch Adams and Urbain Leverrier, the French astronomer. Pluto was discovered in 1930 by Clyde Tombaugh at the Lowell Observatory, Arizona, USA.

The reasons for the delayed discovery of these planets are that they are extremely dim and, being so far away, hard to observe. Even though Uranus can be seen with the unaided eye if you know precisely where to look, Neptune and Pluto cannot.

Uranus is the third of the giant gas planets and shows as a small disc in a powerful telescope. It is crossed by a bright band, which runs from the top to the bottom of the disc and, therefore, lies along the planet's equator. This is due to Uranus being tilted over so that its axis of rotation lies at an angle of some 97.9°, in almost the same plane as the planet's orbit around the Sun.

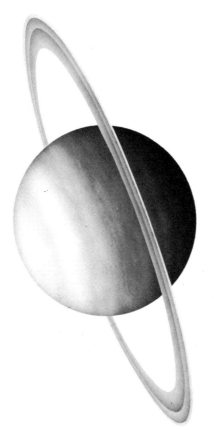

The ring system of Uranus, *drawing left*, is visible only in a large telescope.

An artist's drawing of Neptune, *above*, and its satellite Titan.

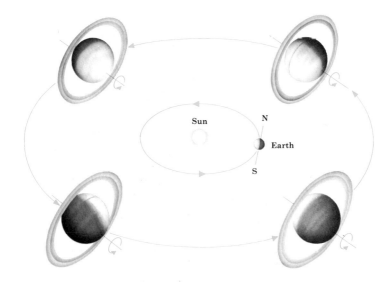

Neptune
Uranus
Pluto
Earth

Uranus has five moons, all of which are too dim to be seen in the ordinary amateur's telescope. In addition, Uranus, like Saturn, has a ring system, containing nine rings in all. This was discovered in 1977, when studies of Uranus as it occulted, or passed in front of, a dim star visible only in a telescope, showed that the star 'winked' before and after occultation, indicating the presence of a system of rings surrounding the planet.

Neptune is the outermost of the gas giants but is visible only in a telescope despite its considerable size. In a large instrument it shows up as a disc, and its axis of rotation is inclined at an angle of 28.8°. Like Uranus, it is completely covered in cloud.

In the coldest regions of the Solar System lies Pluto. To see it at all, you will need a sophisticated, 25-cm (10-in) amateur telescope, and its single moon – Charon – will not be visible. The eccentricity of Pluto's orbit brings the planet at perihelion inside the orbit of Neptune. One theory to account for Pluto's presence beyond the four gas giants is that it is an escaped satellite of Neptune, but this is unlikely. It may, however, be part of an outer belt of asteroids.

Sun

N

Earth

S

Photographs of Pluto, *left and centre left,* taken at the Lowell Observatory in Arizona, show the motion of the planet that led to its discovery by Clyde Tombaugh in 1930. Clues to the correct area of the sky for study came from distortions in the orbit of Uranus. The motion of Uranus in its orbit, *above,* shows that the planet's equator is tilted at 97.9° to its orbit. This means that the rings are never edge-on to an observer on Earth.

The Moon/1

The Moon is the Earth's nearest neighbour in space, orbiting the Earth at an average distance of 384,400 km (238,866 mls). It completes one orbit against the stars (the sidereal month) every 27 days, 7 hours, 43 minutes. It does not shine by its own light, but reflects the light of the Sun and, as a result, shows phases. This is clearly demonstrated by the thin crescent visible soon after New Moon, when the dark face of the Moon is lit brightly enough by sunlight reflected from the Earth for it to be discernible.

The synodic month (from New Moon to New Moon) is 29 days, 12 hours, 44 minutes. Because the diameter of the Moon is only 0.2725 that of the Earth, its gravitational field is less than one-fifth (0.17) of the Earth's. Small as this is, it does exert a pull on the Earth, the most noticeable effect being its influence on the tides.

The Moon's orbit does not coincide exactly with the apparent orbit of the Sun in the sky: they are tilted at an angle of some $5°$ to one another. As a result, on certain occasions the positions of the Sun and Moon coincide, and at times they can lie directly opposite each other. In the first instance, the result is an eclipse of the

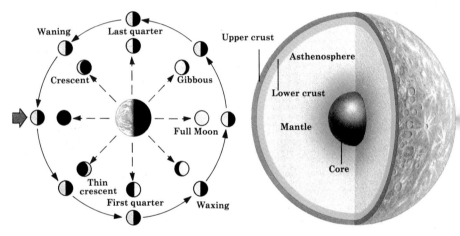

The Moon is illuminated, *outer circle above,* by sunlight from the left. Viewed from Earth, *inner circle,* the phases of the Moon, *below,* can be seen.

The interior of the Moon is totally different to that of the Earth. The outer surface, or regolith, covers a thicker crust and mantle. Below this are the asthenosphere and core.

Phases of the Moon

Sun, with the Moon blocking out some or all of the Sun's light. When the Sun and Moon are opposite each other in the sky, the Moon passes into the shadow cast by the Earth, with a resultant lunar eclipse.

Both lunar and solar eclipses are well worth observing. Lunar eclipses can be seen over large tracts of the Earth's surface, but total solar eclipses, when the whole of the Sun's disc is obscured, are visible from only a small area. A total eclipse is an awe-inspiring sight; the Sun's light goes out, to be replaced by the dim light of its outer atmosphere, or corona,

stars become visible, birds go to roost, and the air temperature drops.

A glance at the Moon on a clear evening will demonstrate that there are features of some kind on its surface, but only a pair of binoculars or a telescope will allow you to see them in detail. Here a telescope, even a small one, is better than binoculars, and the larger the instrument, the more you can see. Powerful telescopes can distinguish many thousands of craters.

The first thing you will notice with binoculars or a telescope is how sharp and clear the features are. This is because the Moon has no atmosphere.

▷

Earth and Moon from space

The Moon/2

American astronaut Buzz Aldrin, the second man to stand on the Moon, accomplished his feat on 20 July 1969, just 18 minutes after Neil Armstrong had made history as the first man on the Moon. The Moon's landscape, and the permanently black lunar sky are reflected in the visor of Aldrin's space helmet.

A map of the Moon shows the features visible through binoculars or a 7.5-cm (3-in) telescope. They should help you become familiar with the Moon's geography. Because the Moon rotates once every time it orbits the Earth, this is the only side of the Moon visible from Earth. The far, or hidden, side of the Moon has been charted by orbiting spacecraft. It is much like the visible side but has fewer maria or 'seas'.

South

Scheiner

Maginus

Tycho

MARE
HUMORUM

MARE
NUBIUM

Gassendi

Riccioli

OCEANUS

Godin

Agrippa   Triesnecker   Kepler

Copernicus

PROCELLARUM

MARE
VAPORUM

Aristarchus

MARE
SERENITATIS

Posidonius

MARE
IMBRIUM

Eudoxus

Hercules   Plato

Aristoteles

MARE FRIGORIS

North

The Moon/3

You can check this by watching the edge or limb of the Moon when it passes in front of, or occults, a star; the star remains undimmed until the moment the Moon occults it. If there were any atmosphere, this would not happen; the star's light would fade away slowly. You do not need a large telescope for observing occultations – 7.5 cm (3 in) is sufficient – but timing such occultations is important in checking the Moon's position.

Most noticeable of the lunar features are the craters and the great flat plains, known as maria (meaning seas), but there is no water on the Moon as astronomers once thought. These maria are frequently fringed by mountain ranges. The craters vary greatly in size; they have mountainous rims, with floors lower than the surrounding surface and often mountains in the middle.

Some, like the crater Tycho, have 'rays' formed of rocky material radiating from them. There are also many smaller features, such as crevices and escarpments, to be seen.

Some observable features of the lunar landscape are, however, more ephemeral – these are known as Transient Lunar Phenomena and take the

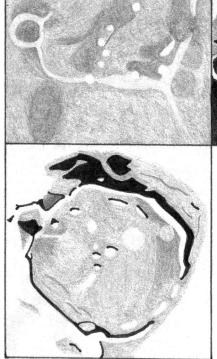

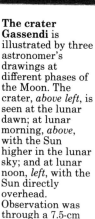

The crater Gassendi is illustrated by three astronomer's drawings at different phases of the Moon. The crater, *above left*, is seen at the lunar dawn; at lunar morning, *above*, with the Sun higher in the lunar sky; and at lunar noon, *left*, with the Sun directly overhead. Observation was through a 7.5-cm (3-in) refractor. Gassendi lies in the WSW region of the Moon on the northern edge of the Mare Humorum. The same crater was photographed, *opposite*, by Lunar Orbiter II in 1966.

form of colorations, brightenings or slight obscuring of the surface detail. Before you can recognize such brief changes in surface detail, you must become familiar with lunar features under a variety of different light conditions. Depending on how the light from the Sun strikes the surface of the Moon, features can appear dramatically different. One likely explanation for Transient Lunar Phenomena is that they are some sort of gas release from inside the Moon.

Because the Moon rotates only once on its axis for every complete orbit of the Earth, it always keeps the same face toward us. Yet details on this face change continually as the angle of the Sun's light alters. The best time to observe detail is not at Full Moon, when the even illumination tends to make features appear flat, but at other times when the sunlight shines more obliquely. Illusions of depth are created as slanting sunlight illuminates the towering sides of mountains but leaves crater floors in inky blackness. Fine sights can also be observed close to the terminator (the edge between lit and unlit sides), with the peaks of the mountains just catching the Sun's rays.

Earth and Moon both cast shadows in space. When the Moon moves into the Earth's shadow, a lunar eclipse results, which can be seen from anywhere on the night-side of the Earth. When the Moon casts a shadow, a solar eclipse occurs, but since this shadow covers only a small portion of the Earth's surface, the eclipse is visible only from a local area. However, as both Earth and Moon are orbiting in space, the eclipse shadow moves to give a slightly wider area of observation than shown in the diagram, *right*.

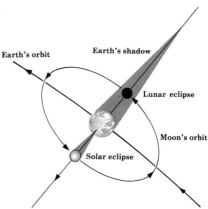

The stars/1

Glance up at the sky on a clear night, and you will see that the stars vary widely in their apparent size and brightness. (Brightness is discussed in detail on pp 138–41.) Stars also vary in colour, but colour differences can be detected by the unaided eye for only a few of the brighter stars – Betelgeuse and Rigel in the constellation of Orion, for example, or Castor and Pollux in the constellation of Gemini.

If you have a telescope, you will be able to distinguish colour differences in more stars, but again just in the relatively bright ones, since the eye can detect colour only when it is viewed in fairly strong light. Test this by taking a variety of differently coloured objects outside and looking at them by moonlight. Although moonlight is reflected sunlight, it is still too dim for accurate colour identification.

Detecting the colours of stars visually is somewhat imprecise, and astronomers specify star colours exactly by analyzing the light that a star emits. This they do by splitting up the light, using a prism or a grating (a surface that has thousands of etched lines per centimetre). Both these devices break up light into a spectrum, or bands of

α	alpha	**Stars** are often designated with Greek letters rather than names. On the pages that follow these letters are used extensively. The chart shows the Greek alphabet with its classical letters and their names.
β	beta	
γ	gamma	
δ	delta	
ϵ	epsilon	
ζ	zeta	
η	eta	
θ	theta	
ι	iota	
κ	kappa	
λ	lambda	
μ	mu	
ν	nu	
ξ	xi	
o	omicron	
π	pi	
ρ	rho	
σ	sigma	
τ	tau	
υ	upsilon	
ϕ	phi	
χ	chi	
ψ	psi	
ω	omega	

Every star exerts a gravitational pull. It can be thought of as making a 'dent' in the surrounding space, *below left*. This means that

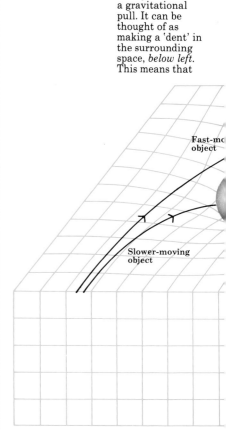

Fast-moving object

Slower-moving object

colour, such as can be seen when sunlight strikes the prism-shaped edge of a mirror or the surface of a video disc, for example.

Breaking down starlight into a spectrum allows astronomers to discover the temperature, true brightness and chemical composition of the star under investigation. This is done by studying the brightness of various parts of the spectrum, as well as the dark lines that cross its brightly coloured background.

The dark lines in the spectrum are important because they indicate whether a star is moving toward or away from Earth. This is the only way to obtain such information, since stars, even in the largest telescopes available, only ever appear as tiny points of light. It is not possible actually to witness an apparent change in star size because the distances involved are so immense. Astronomers now know the sizes of most stars, due not to direct observation but to calculations based on temperature, brightness and spectral emissions.

That these three factors are connected was first demonstrated in 1914 by the Danish astronomer Ejnar Hertzsprung and the American Henry

▷

any body that goes near will fall toward it. The sort of dent a star makes will depend on how massive it is. The heavier the star, the steeper and more widespread the dent. White dwarfs and neutron stars are massive and make broad, steep dents. Black holes exert such a gravitational pull that not even light can escape from them. Their dents are the steepest of all, *below right*. In them, space 'folds over' so that material falling into them is never seen again. Bodies near the dents of such stars must be moving very fast if they are not to fall on to the star.

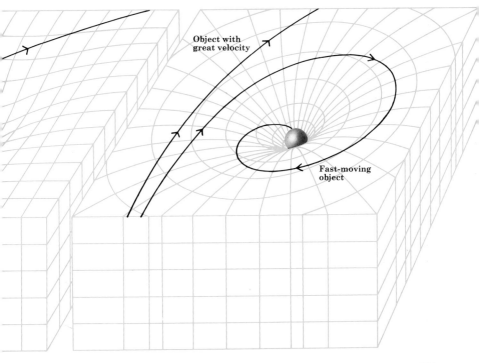

Object with great velocity

Fast-moving object

The stars/2

Norris Russell in the famous H-R, or Hertzsprung-Russell, diagram. From this diagram you can see that the hottest stars – the blue-white ones – are the brightest, while the many red stars are, in comparison, extremely dim. There is, in fact, a 'main sequence' of stars running from cool red to hot blue-white. Our own Sun lies on this sequence. Nevertheless, some cool red stars are extremely bright and there are some hot white stars that are rather dim.

It has also been found that the H-R diagram displays the life sequence of stars. Stars, which are composed mainly of hydrogen gas, usually begin their evolution as small, dim, cool, red bodies. They shine because, at their centres, the hydrogen atoms are broken down into helium – a star is, in fact, like a gigantic hydrogen bomb. It is estimated, for instance, that our own Sun consumes approximately a million tons of its own matter every second.

In the next evolutionary stage, stars gradually heat up and change colour. But they stay on the main sequence for only a given time and when much of their hydrogen has been changed into helium, stars expand and cool.

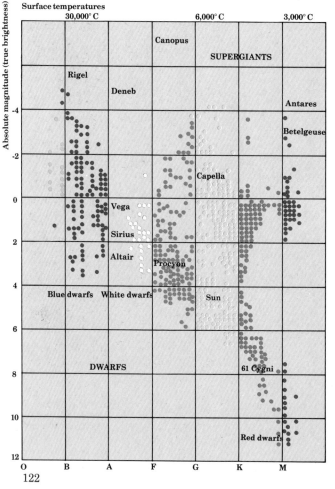

The Hertzsprung-Russell diagram, *left*, classifies the stars according to their spectral type, or colour (left to right), and their absolute magnitude, or true brightness (top to bottom). From the diagram, some characteristics are immediately apparent – the blue stars are all bright ones and red stars are divided into two classes (bright giants and dim dwarfs). There is also a main sequence of stars running from red dwarfs (lower right) to bright blue (upper left). Our Sun lies on this sequence. The diagram also delineates temperatures of the stars, from hot blue ones to cool red ones.

When our Sun reaches this stage, in approximately 5,500 million years, it will be sufficiently large to engulf Mercury and Venus and probably the Earth as well. At this point stars are known as red giants or supergiants. Finally, they shrink until their centres become fantastically dense – a teaspoonful weighing tons – and their outer regions hot; they are then known as white dwarfs.

If a star is really massive when it first forms, it will quickly reach a hot stage and become one of the hot white or blue-white types. At the end of its life, it will collapse to become far denser than a white dwarf. It will probably turn into a neutron star, where a teaspoonful would weigh millions of tons. But if a star is extraordinarily big to begin with, it will eventually collapse until its atoms are completely crushed; then, because of the immense gravitational field associated with it, it emits no light, heat or any other kind of radiation. It becomes what is known as a black hole, drawing everything nearby into it, just as if the material really were falling into a hole in space. Even beams of light passing close to its gravitational field are deflected.

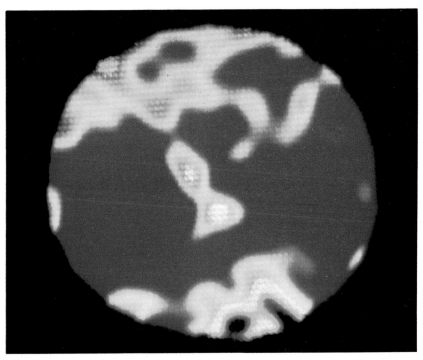

A speckle interferometer on a telescope at Kitt Peak National Observatory in the USA was used to take the unusual photograph, *above,* of α (alpha) Orionis (Betelgeuse). This instrument builds up an image of a star that can be processed by a computer to provide information on its surface and its appearance. The colours shown are false, but they do indicate that Betelgeuse displays patches which are cooler areas on the stellar surface.

The Galaxy/1

For a skywatcher appearances can be deceptive. The stars appear as if they are fixed to the inside of a dome or a sphere, and indeed this is how early astronomers believed them to be: fixed in the dome of space. Not until 1576 were ideas about an infinite universe of stars published by the Englishman Thomas Digges. Yet even when this idea came to be accepted, astronomers still thought the stars to be fixed in position.

Astronomers believed this because this is how they appear to be, unmoving, settled in their specific places in the many constellations. Every generation has observed them to be the same. Yet they are moving; it is only their great distance that makes it impossible to detect this movement with the unaided eye. It was Edmond Halley who first detected such motions in 1717, yet it was not until some 170 years later that they were studied in detail.

Each star has an individual motion. In addition, all are orbiting around the centre of a vast island of stars, dust and gas, known as a galaxy. It has been estimated that our Sun takes approximately 225 million years to complete one full galactic orbit. You

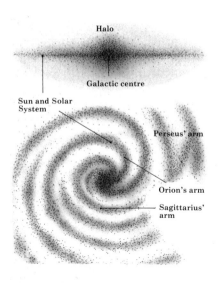

Halo

Galactic centre

Sun and Solar System

Perseus' arm

Orion's arm

Sagittarius' arm

+60°
+40°
+20°
0°
−20°
−40°
−60°

PHOTOGRAPHIC MAGNITUDES

The edge-on view of the Galaxy, *top*, does not show the spiral arms. These can be seen only when looking down on the Galaxy, *above*. The missing part is on the far side of galactic centre, invisible from Earth.

A composite of many photographs was used by the Lund Observatory in Sweden to create this view, similar to a map projection. It shows the concentration of stars and star fields in the plane of the Galaxy.

can see part of our Galaxy any night when the stars are visible. It appears as a hazy band of light, called the Milky Way, which stretches across the sky from horizon to horizon, and is visible from both northern and southern hemispheres of the Earth.

Though the Milky Way has been a feature of the night sky remarked upon from ancient times, until 1609 its true nature always posed a question. In that year, the Italian astronomer Galileo observed the Milky Way with his telescope and discovered that it is composed of a myriad of individual stars. Even so, the fact that it is a

star-island remained unknown until the 1780s, when the British astronomer William Herschel, an observer of the greatest skill, using what were then the world's largest telescopes, arrived at the idea. However, he thought it was shaped like an elongated box, with our Sun located close to the centre.

It was not until the 1920s that the true nature of the Milky Way was finally determined, using improved telescopes fitted not only with cameras but also with spectroscopes for analysing starlight. It is now known that the Milky Way system – the ▷

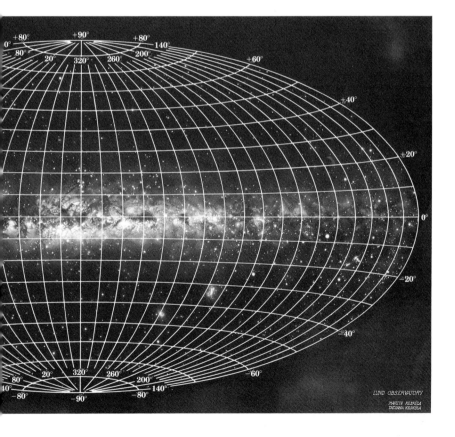

The Galaxy/2

Galaxy – is a vast, flattened disc with a large central bulge. The use of radio-telescopes has also shown that the Galaxy is embedded in a corona of dust and gas.

The extent of the optically visible system is enormous, and to attempt to describe it in terms of kilometres or miles proves to be inadequate, as, indeed, it is for describing the distances of even nearby stars. There is a need, therefore, to adopt much larger units of distance.

The distance unit normally used in popular literature on astronomy is the light-year. A length, not a unit of time, it is the distance over which light travels in one year. Since light travels at about 300,000 km/sec (186,000 mls/sec), one light-year amounts to no fewer than 9.46 million million km (5.87 million million mls), or more than 62,000 times the distance from Earth to the Sun.

To determine stellar distances, astronomers begin by measuring those of the nearer stars. To do this, they observe such stars from each side of the Earth's orbit, in other words, at six-monthly intervals. This gives them the parallax (the apparent change of position when an object is viewed from two different positions) of the star and involves measuring tiny angles, typically fractions of an arc second (1 arc second = 1/3,600 of a degree).

Because such distances involve making angular measurements, astronomers find it more convenient to use a unit known as the parsec – the distance at which a star would have a parallax of one arc second – equivalent to a distance of 3.26 light-years. The Sun, of course, is nearer than this, but no other star has a parallax as small as one parsec. The terms kiloparsec (kpc) and megaparsec (mpc) are used for one thousand and one million parsecs respectively.

Using these scales, astronomers have found that the diameter of the

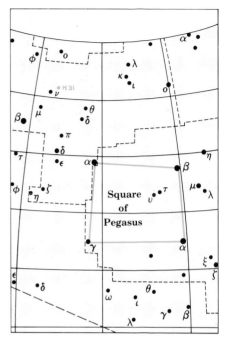

The Square of Pegasus and part of the Andromeda constellation are shown in the diagram, *above*. The vast galaxy M31, sometimes known as the Great Nebula in Andromeda, lies above and a little to the right of ν (nu) Andromedae.

The Andromeda galaxy, M31, with its two companion galaxies, are revealed in this photograph. It was taken with the 1.3-m (48-in) Schmidt telescope at the Hale Observatory, California. M31 is a member of the Local Group of galaxies, which includes our own. The separate stars, visible like a gauze curtain over the photograph, are all stars in our Galaxy. An observer must look beyond them to see M31.

▷

The Galaxy/3

disc of our Galaxy is some 100,000 light-years (30 kpc), and the thickness of the central bulge is about 10,000 light-years (3 kpc). Its corona is vast, enveloping the visible Galaxy out to a distance of at least 200,000, and possibly 326,000, light-years (60 or 100 kpc). Inside this corona, and surrounding the disc and central bulge, are hundreds of thousands of globular clusters of stars (see pp 160–5). Our Sun and its Solar System lie in the main plane of the Galaxy, some two-thirds of the way toward the edge of the disc.

The fact that the Earth is positioned in the plane of the Galaxy is the reason why the Milky Way appears to stretch over the entire sky. When you look in the direction of the Milky Way, you are sighting along the plane of the Galaxy; almost all the other separate stars you see in the sky lie above or below this plane. The centre of the Galaxy lies in the general direction of Sagittarius (the Archer), hence the dense clouds of stars in that region of the sky. You cannot observe the centre of the Galaxy itself for there is too much dust and gas in the way.

During this century it has been discovered that our Galaxy is not alone;

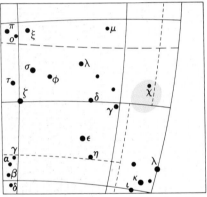

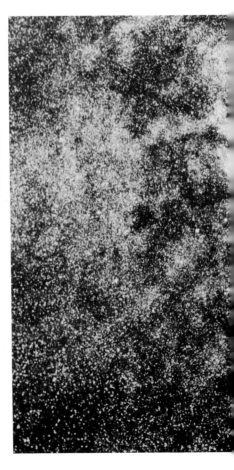

The position of the star cloud in Sagittarius can be represented diagramatically, *above*. The star cloud is also represented photographically, *right*, looking toward the centre of our Galaxy. The centre is, however, invisible because of the stars, dust, dark and bright nebulous gas clouds that lie in the way. Each dot in the photograph is an image of a distant star. Some are a little dimmer, but most are as bright as, or brighter than, our own Sun. And some may also have planetary systems as our Sun does. The region does not look quite like this through a telescope, but it is rich in stars and well worth studying.

large telescopes have been used to photograph hundreds of thousands of other galaxies. These are of two types – elliptical and spiral. Elliptical galaxies are, as their name implies, elliptically shaped islands of stars, with little if any gas, typically thousands to several hundreds of thousands of light-years across. Some, it is true, appear to have a spherical shape, but this is probably simply an optical effect due to the angle from which they are viewed.

Spiral galaxies, on the other hand, contain not only stars but much gas and dust as well. They have a central bulge with a surrounding disc that displays spiral arms, and radiotelescopes have helped to confirm that our Galaxy is just such a spiral. The size of the spirals is similar to that of the ellipticals; both are massive, and they range from about 2 million million to 3 million million times the mass of the Sun.

The stars in elliptical galaxies are all fairly old (about 10,000 million years) and are known as Population II stars. Those in spiral galaxies show wider variety: the stars in the region of the central bulge are Population II, but in the spiral arms they are

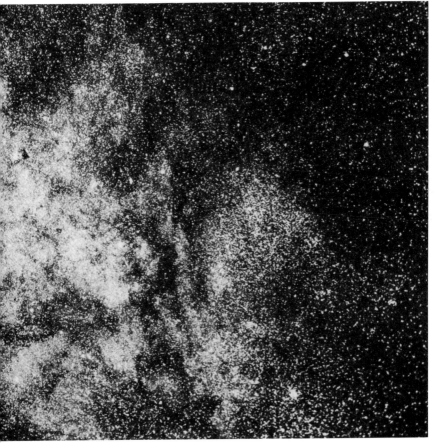

The Galaxy/4

younger, probably not more than 4,000 million years old, and there are many hot, blue supergiants – these are the Population I stars.

Galaxies are, for the most part, gathered into clusters. Our Galaxy belongs to such a cluster, known as the Local Group, which contains at least 21 members. One of these is the Andromeda galaxy, commonly known by its catalogue number, M31. You can see it with the unaided eye as a hazy patch of light some distance north of β (beta) Andromedae. If you look at it in an ordinary telescope of about 15 cm (6 in), however, it is a disappointing sight compared with its well-known photographic image. This is because it is so distant – some 2.2 million light-years away; it is, in fact, the most distant object that can be seen with the unaided eye. As with a number of other galaxies, including our own, it is attended by companions. If you observe from the southern hemisphere, you can see our two companion galaxies, the Large and Small Magellanic Clouds.

By and large, the distances of galaxies are so great – some have been found thousands of millions of light-years away in space – that they are

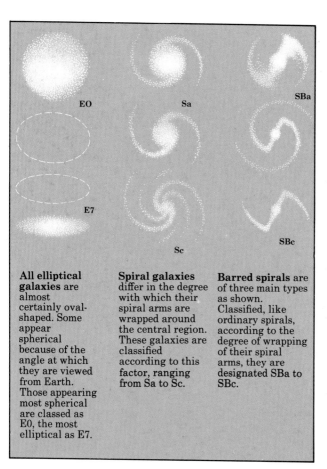

All elliptical galaxies are almost certainly oval-shaped. Some appear spherical because of the angle at which they are viewed from Earth. Those appearing most spherical are classed as E0, the most elliptical as E7.

Spiral galaxies differ in the degree with which their spiral arms are wrapped around the central region. These galaxies are classified according to this factor, ranging from Sa to Sc.

Barred spirals are of three main types as shown. Classified, like ordinary spirals, according to the degree of wrapping of their spiral arms, they are designated SBa to SBc.

unsuitable objects for viewing in all but a few large amateur telescopes of 25 cm (10 in) and upward.

By observing far into space, evidence has been found that in the distant past some galaxies were highly explosive, radiating vast amounts of energy. Indeed, such bodies are what the professional astronomer refers to as quasi-stellar objects, or quasars. But not only quasars are active; some nearby galaxies are extremely active and undergo explosions, ejecting vast amounts of material into space; still others have bright, active cores.

One peculiarity of all galaxies is that the lines in their spectra are all shifted toward the red end. This redshift means that they are moving away from us, some at quite astonishing speeds. One galaxy, referred to simply by its catalogue number, 3C-295, is receding at a velocity approaching half the speed of light. This type of movement has given confirmation to the idea that the universe is expanding. It has also led astronomers to the Big Bang theory – the view that all the heavenly bodies that can now be seen in the night sky began from an extremely compact point some 18,000 million years ago.

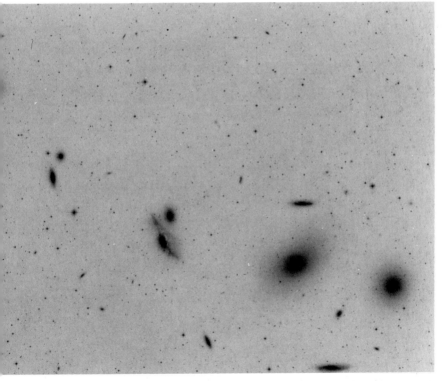

A cluster of galaxies in the constellation of Virgo can be seen, *above*. It is possible to discern elliptical galaxies; these are round and 'fuzzy'. The galaxy in the centre is NCG 4435. The average distance of the cluster from Earth is 70 million light-years.

The constellations/1

If you want to make more than a cursory examination of the heavens, it is necessary to classify the stars in some way and to find a means of referring to at least the brightest ones. All civilizations seem to have devised methods of doing this, usually by grouping the stars into patterns. In the past, these patterns were linked either to the gods, goddesses and legendary heroes of the particular culture or to characters associated directly with the more notable deities.

At such times, the study of the heavens could most properly be described as astrology. The skies were scanned for indications of the fates of rulers and the blessings or punishments meted out by the gods. At a time when stars and planets were thought to be comparatively close, and no ideas had been formulated about their physical nature, this was plausible.

With our present knowledge of the nature of the planets, stars and constellations, as well as of other celestial events, such as the appearance of comets or a new star (a nova, or supernova), these beliefs are no longer tenable. Even so, we have astrology to thank for its stimulus to observation of the skies; rulers wanted to know

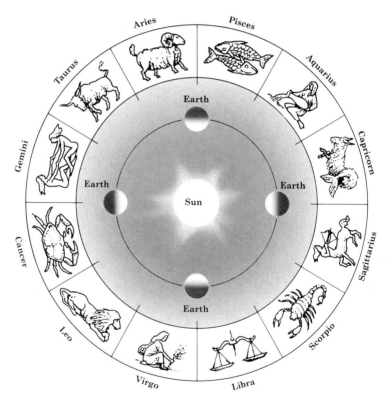

Sun, Moon and planets, including the Earth, *above*, appear to move through the zodiacal constellations as they circle the sky.

A chart of the circumpolar stars from *Harmonia Macrocosmica* by Andreas Cellarius, published in 1660, *opposite*, shows the brighter stars and the constellation figures.

both the opinion of the gods and what the future held.

The constellations known today (see pp 142–59) all have Latin names, although they do not derive from ancient Rome. They were recognized, for the most part, by the old Persian civilization centred on Babylon and date from the fifth century BC. Taurus (the Bull), Gemini (the Twins), Scorpio (the Scorpion), Sagittarius (the Archer), Capricornus (the Sea Goat), Aquila (the Eagle), Leo (the Lion), Lupus (the Wolf) and Corvus (the Raven) were all Babylonian constellations. But some that the Persians recognized were rearranged by the Greeks, their Panther, Goat and Bowl becoming our Cygnus (the Swan), Lyra (the Harp) and Auriga (the Charioteer).

More recent astronomers did not draw on the legend and rich imagery of these ancient civilizations when it came to naming some of the constellations of the southern skies. Telescopium (the Telescope) and Microscopium (the Microscope) are two examples displaying a more prosaic nomenclature.

Although many constellation names remain the same, their actual extent

▷

The constellations/2

may well be different, since present-day astronomers use constellation boundaries laid down in detail by the International Astronomical Union.

The IAU boundaries are laid out in coordinates which are used for defining precisely positions in space. They are, in other words, the equivalent on the sky, or celestial sphere, of latitude and longitude on Earth. The main reference circle is the celestial equator, which lies exactly between the north celestial pole (marked approximately by Polaris) and the south celestial pole for which there is no Polaris equivalent. Crossing this circle at an angle of 23.5° (the angle of axial inclination of the Earth) is the path of the Sun, known as the ecliptic.

At the crossing points, night and day are of equal length. The crossing points move slowly along the celestial equator. This is known as the 'precession of the equinoxes'; it amounts to only 50 arc seconds per year. The starting point of the equivalent of terrestrial longitude is the spring (vernal) equinox, when the Sun moves to the north of the celestial equator. It is measured eastward along the celestial equator, and is not called longitude but right ascension (RA).

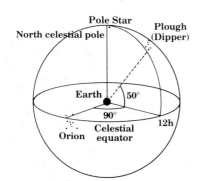

The celestial sphere, *left,* shows both right ascension and declination. Right ascension is measured from the First Point of Aries (the vernal equinox), where the ecliptic crosses the celestial equator.

Different stellar magnitudes are clearly shown in the photograph of the Plough (Dipper), *below.* The picture was taken using fast, ISO 400 colour film and a driven camera (see pp 188–91).

The units of right ascension are hours, minutes and seconds of time, not the terrestrial system of degrees, minutes and seconds of arc (but note: 1 hour of right ascension equals 15°).

The equivalent of terrestrial latitude is called declination (dec). Declinations north of the celestial equator are counted as positive and those to the south are counted as negative. Declinations are measured in degrees. Thus the coordinates of Betelgeuse are: RA 5h 55m 10.2s, dec $+7°$ 24' 26''; and for Rigel: RA 5h 14m 32.2s, dec $-8°$ 12' 06''. When using celestial coordinates, 1 hour is divided up into 60 minutes, and each minute into 60 equal seconds.

Some constellations are not at all easy to recognize and the only way to learn their positions is to become familiar with the night sky. For this, you need no telescope, although a planisphere (a circular star chart rotating under a mask that shows the night sky) is helpful.

For northern skies, begin with the Plough, or Dipper. These seven stars are always above the horizon unless you are close to the Equator, and the constellation is easy to detect. An imaginary line drawn through its two

▷

The Orion constellation, *right,* is one of the easiest to recognize. This photograph not only indicates the general pattern of the whole constellation but also the nebulous area in the region of Orion's sword, which lies directly below the belt.

The constellations/3

right-hand stars – the Pointers – leads to Polaris (the Pole Star), which is at the end of the tail of Ursa Minor (the Little Bear). The stars of the Plough are part of the constellation of Ursa Major (the Great Bear), but the remaining stars are dimmer and not so easy to identify.

On the opposite side of Polaris is Cepheus, and at the same distance but over to the right is the W of stars known as Cassiopeia. Keep in mind, though, that the Earth rotates on its axis, so the constellations appear to circle around Polaris. As a result, you will see them in all positions and

orientations – on their sides or even upside-down – so the W of Cassiopeia may appear like an M, for example.

Some other constellations are also easy to recognize and can be used as guides to the more difficult ones. Such are the Square of Pegasus, Gemini and Taurus (which you can recognize by its bright red star, Aldebaran, and the tiny cluster of stars known as the Pleiades). One of the easiest constellations to identify is Orion (the Hunter), with its two bright stars, Betelgeuse and Rigel, its diagonal belt of stars in the middle and the sword hanging downward from the centre of

A **detail** from a British Astronomical Association star chart drawn by the Dutch amateur astronomer Wil Tirion, *right*, shows some southern hemisphere stars. The coordinates of right ascension and declination have been corrected for precession – a movement of the Earth's axis – so the stars are drawn as they will appear in the year 2000. The Southern Cross is present, as is the larger Magellanic Cloud, which is a companion galaxy to our own.

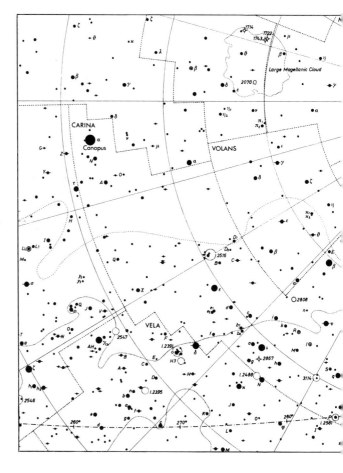

the belt. Using the line of Orion's belt as a guide, look downward and you come across Sirius in Canis Major (the Great Dog); sight upward along the belt and you find Aldebaran in Taurus. Unfortunately, Orion is not visible in northern skies during the summer.

For observers in the southern hemisphere, the circumpolar constellation, Crux (the Southern Cross) is always visible. The vertical bar of the cross points toward the south celestial pole. Other notable constellations of the southern hemisphere are Carina (the Keel), Scorpius (the Scorpion), Centaurus (the Centaur), Canis Major (the Great Dog), Puppis (the Poop) and Vela (the Sails).

That a certain collection of stars forms a constellation is purely an optical effect; they are not connected, as astrologers believed. Taking Orion as an example, its stars are all at different distances from Earth. From Earth to Betelgeuse is almost as far as from Betelgeuse to Rigel, and the left-hand star in the belt is nearly 9 times nearer the Earth than Betelgeuse, while the centre one is 6 times nearer, and the right-hand one is more than twice the distance of Rigel. That we see them as a group is a line-of-sight effect.

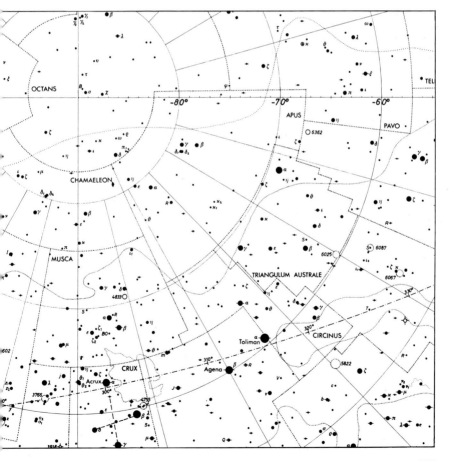

Star magnitudes/1

Looking up at the night sky, three things become obvious. The first is that the stars twinkle, the second is that it is easy for us to arrange some into recognizable patterns, and the third is that not all the stars look the same. Some are bright, some less bright, and some almost too dim to see with the unaided eye.

Because of the brightness differences, astronomers have had to devise some means of expressing the variation precisely. The attempt to classify such differences began some two thousand years ago when the ancient Greek astronomer Hipparchos

produced a star catalogue in which he introduced the idea of importance, or magnitude, for the stars he listed.

Hipparchos thought the brightest stars to be of first importance and termed them magnitude 1. Those somewhat dimmer he considered of second importance, or magnitude 2, and so on. In the event he divided the stars visible to the eye into six classes of magnitude. Although he did not know it, his six divisions were based on the way the human eye recognizes a brightness difference where one object seems half as bright as another. It has subsequently been found also that

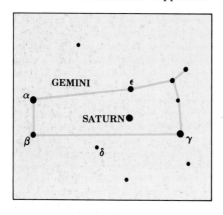

A star map of the constellation Gemini (the Twins) is shown, *above*. The two bright stars Castor and Pollux are clearly visible. Castor has an apparent magnitude of +1.58 and is blue-white in colour. Pollux is slightly brighter at +1.14 but appears redder. Ancient records indicate that Castor may once have been brighter than Pollux.

Gemini, shown in the photograph, *right*, contains the two first-magnitude stars Castor and Pollux. This fine zodiacal constellation is one of the easiest to recognize.

Hipparchos' six magnitudes give a difference between magnitudes 1 and 6 of 100 times.

In the intervening two thousand years the subject of magnitudes has been much refined. Yet astronomers still use the same basic system, which is really a measure of dimness – the higher the magnitude number, the dimmer the star. To be more precise, a star of magnitude 2 is 2.51 times dimmer than one of magnitude 1, and a star of magnitude 3 is 2.51 times dimmer than one of magnitude 2. For extremely bright stars, the planets and the Sun and Moon, it is necessary to introduce magnitude zero and even negative magnitudes: Sirius is described as having a magnitude of −1.46, while the Sun is −26.85.

Most stars do not have names, but even if they do, you will still find that the brightest are also designated by a Greek letter, starting with α (alpha) for the brightest, β (beta) for the next brightest, and so on down the alphabet. Since there are more stars to be designated than there are Greek letters, each of the remaining stars is assigned a number.

Stars and other celestial objects often have catalogue numbers as well.

▷

Star magnitudes/2

Thus Sirius is also α (alpha) Canis Majoris; the second magnitude star in Orion's left foot is κ (kappa) Orionis, while the fifth magnitude star just above it to the left is 55 Orionis.

All these brightnesses describe the way that stars appear to observers on Earth. They are, therefore, apparent magnitudes and do not describe the true brightness of the star. Thus Sirius has an apparent magnitude of −1.4 and the Sun −26.85, a difference of 25.4 magnitudes. This translates into a difference of brightness of 14 thousand million times. Certainly this is how Sirius and the Sun appear from Earth, yet it is known that Sirius is nearly 25 times brighter and appears so dim only because of its remoteness.

To overcome this type of confusion, astronomers also use the term absolute magnitude. This is the magnitude stars would have if they were all fixed at a specific distance from Earth. The arbitrary distance chosen is a convenient one for astronomers – 10 parsecs (equivalent to 32.6 light-years). Using this system, the absolute magnitude of Sirius is +1.41 and that of the Sun only +4.83, which shows that, in reality, the Sun is almost 25 times dimmer than Sirius.

A star's annual parallax – the apparent shift of a star against the background of more distant stars – is measured by making observations of the star six months apart, when the Earth is at the opposite side of its orbit, *below*. The degree of annual parallax is half the angle subtended by the Earth's orbit at the distance of the star.

The marked differences in brightness between the various stars in Orion is visible, *right*. Observers in the southern hemisphere will see Orion 'upside-down'.

The constellation of Leo, *far right*, is one of the few zodiacal constellations that has a shape like that of the creature or object it is supposed to depict.

Sun

Earth's orbit

The apparent magnitude of a star depends to some extent upon its colour. This is because the eye is most sensitive to greenish-yellow light, whereas ordinary photographic emulsion (despite being panchromatic) is most sensitive to the blue end of the spectrum and least sensitive to the red end. You can test this by looking at a photograph of Orion, when you will notice that the bright red star Betelgeuse appears much dimmer than the blue-white star Rigel. To the naked eye, however, they both appear extremely bright stars.

Astronomers take account of these differences by specifying precisely the colour system they are using, but this is not important if you simply want to observe the stars through a telescope. Ordinary visual apparent magnitude will also be sufficient for observing the way the amount of light that is emitted by some stars varies. If you intend to make photographic records of stars (see pp 188–91), you must take account of the way photographic emulsion responds to light of different colours.

Another method of classifying stars is by their total radiation output, not just the visible light. This is known as bolometric magnitude.

Parallax

Star watching/1

The charts that follow (pp 144–59) are designed for observation of stars with the naked eye, down to the fifth magnitude. All the important constellations are included, and their relationship to each other in the night sky indicated.

Coordinates given are those of right ascension and declination (based on the celestial equator), which are the equivalents of terrestrial longitude and latitude.

In modern maps, star positions, as well as the positions of other celestial bodies, are always given using right ascension and declination. The coordinates known as celestial longitude and latitude are reserved primarily for computing the positions of the planets. Celestial longitude runs eastward along the ecliptic, and celestial latitude north and south toward the poles of the ecliptic.

Each hour in right ascension is equivalent to 15°. Zero right ascension begins where the celestial equator and the ecliptic intersect – the spring, or vernal, equinox. This is the point at which the Sun crosses the celestial equator and moves into the northern half of the celestial sphere.

Star atlases contain a third set of coordinates (not shown here), which concerns galactic latitude and longitude. These coordinates are based on the Milky Way instead of the ecliptic or the celestial equator, and they are used in studies of the distribution of globular clusters (see pp 160–5) and galaxies (see pp 124–31).

Observers using a telescope or binoculars will need a star atlas that gives more detail than it is possible to show on the small-scale maps reproduced here. One such widely used atlas is *Norton's Star Atlas* published by Gall & Inglis, Edinburgh; another consists of the detailed star charts drawn by Wil Tirion and published by the British Astronomical Association. These charts are also available from Enslow Publishers, New Jersey, USA.

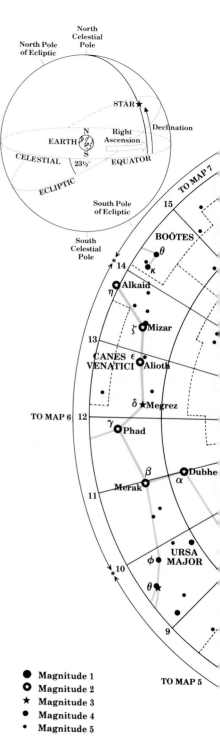

● Magnitude 1
◉ Magnitude 2
★ Magnitude 3
● Magnitude 4
· Magnitude 5

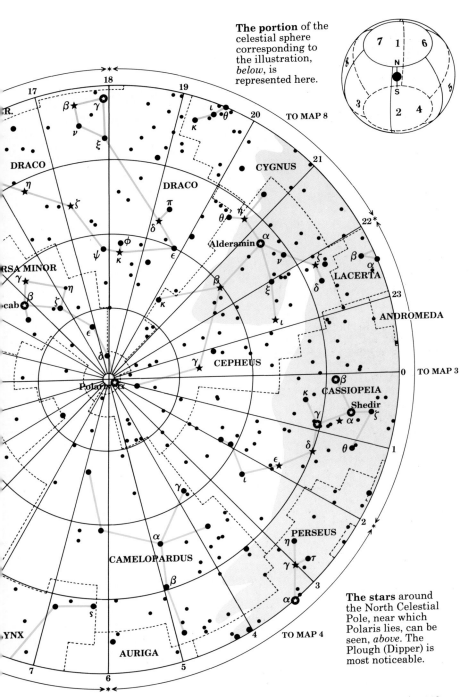

The portion of the celestial sphere corresponding to the illustration, *below*, is represented here.

7 1 6
8
N
3 S
2 4
5

The stars around the North Celestial Pole, near which Polaris lies, can be seen, *above*. The Plough (Dipper) is most noticeable.

DRACO

CYGNUS

DRACO

URSA MINOR

Alderamin

LACERTA

ANDROMEDA

TO MAP 8

TO MAP 3

CEPHEUS

CASSIOPEIA

Shedir

Polaris

CAMELOPARDUS

PERSEUS

LYNX

AURIGA

TO MAP 4

143

Star watching/2

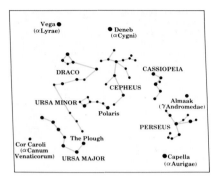

Some notable circumpolar constellations are shown, *left.*

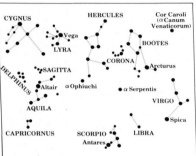

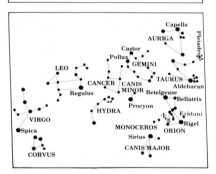

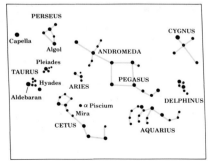

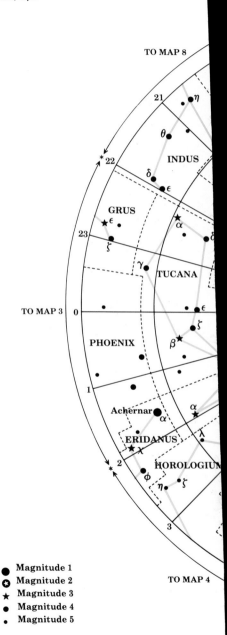

● Magnitude 1
✪ Magnitude 2
★ Magnitude 3
• Magnitude 4
· Magnitude 5

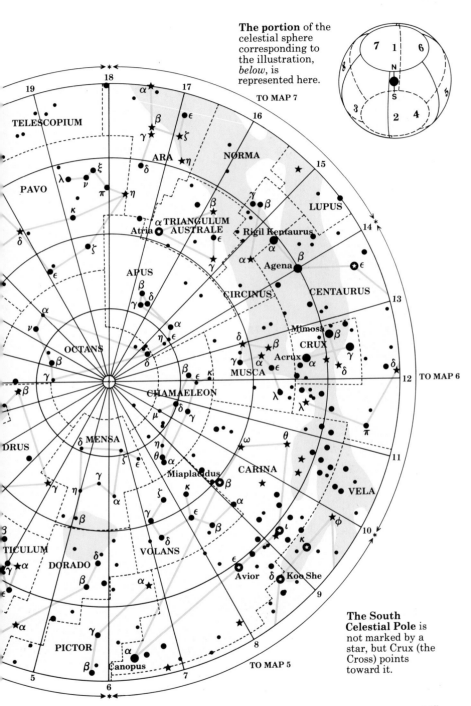

The portion of the celestial sphere corresponding to the illustration, *below*, is represented here.

The South Celestial Pole is not marked by a star, but Crux (the Cross) points toward it.

145

Star watching/3

Looking south in the northern hemisphere (north in the southern), the main constellations visible at midnight on 22 November (10 pm on 22 December and 8 pm on 21 January) are as shown. Notable constellations are the Square of Pegasus and Andromeda.

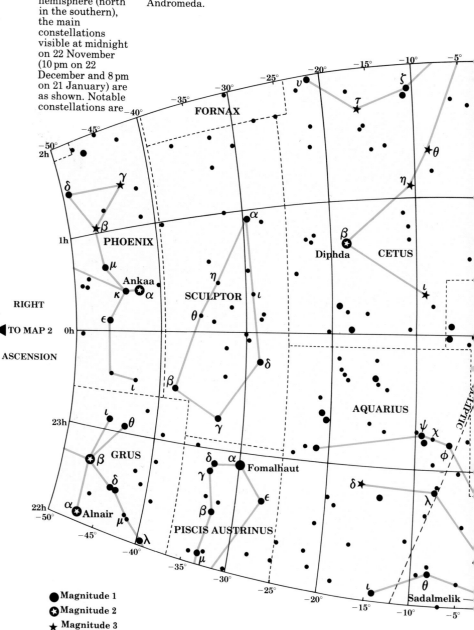

RIGHT

◄ TO MAP 2

ASCENSION

- ● Magnitude 1
- ✪ Magnitude 2
- ★ Magnitude 3
- ● Magnitude 4
- • Magnitude 5

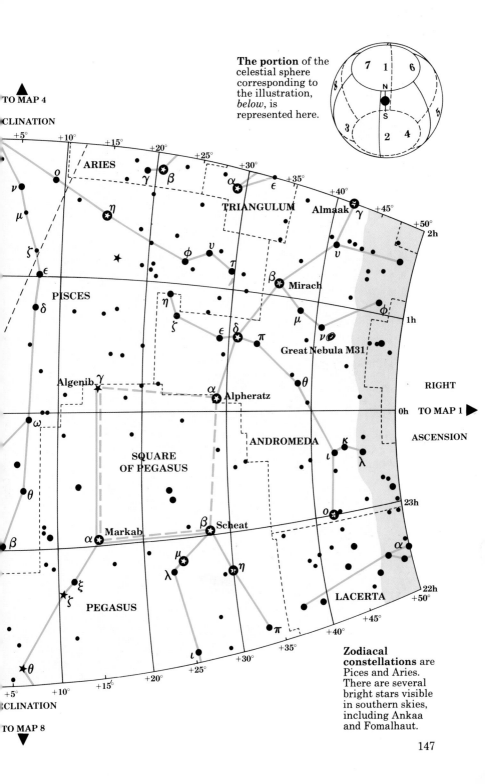

TO MAP 4

The portion of the celestial sphere corresponding to the illustration, *below*, is represented here.

CLINATION

+5°   +10°   +15°   +20°   +25°   +30°   +35°   +40°   +45°   +50°   2h

o

ν

μ

ARIES

γ   β   α   ε

η   Almaak   γ

φ   ν   τ   ν

TRIANGULUM

ζ

ε

PISCES

δ   η   ζ

ε   δ   π

β   Mirach

μ   φ   1h

ν   Great Nebula M31

θ

RIGHT

Algenib   γ   α   Alpheratz

ω   ASCENSION

0h   TO MAP 1 ▶

κ   ANDROMEDA

θ   SQUARE OF PEGASUS   ι   λ

β   o   23h

α   Markab   β   Scheat   α

μ   22h   +50°

λ   η

ξ   LACERTA

ζ   +45°

PEGASUS   +40°

ι   +35°

π   +30°

θ   +25°

+5°   +10°   +15°   +20°

CLINATION

TO MAP 8

Zodiacal constellations are Pices and Aries. There are several bright stars visible in southern skies, including Ankaa and Fomalhaut.

147

Star watching/4

Looking south in the northern hemisphere (north in the southern), the main constellations visible at midnight on 23 November (10 pm on 23 December and 8 pm on 22 January) are as shown. In northern skies, note Perseus and Auriga, with the bright star Capella.

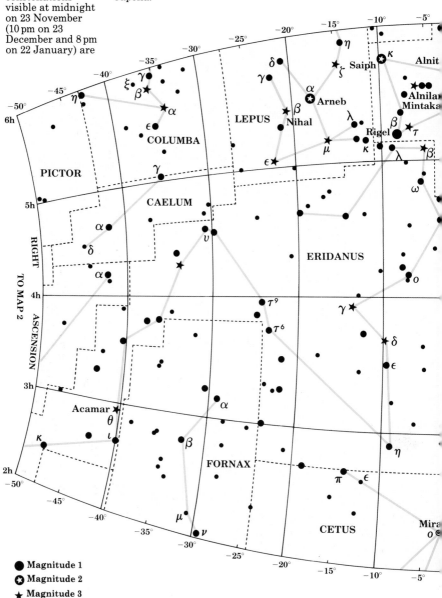

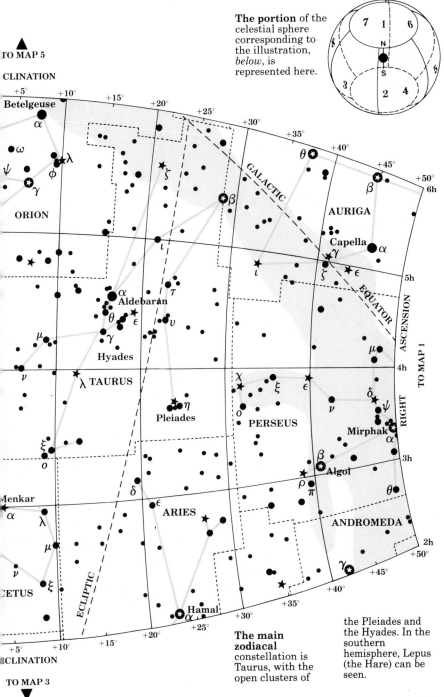

The portion of the
celestial sphere
corresponding to
the illustration,
below, is
represented here.

7 1 6

8

N

S

3 2 4

5

CLINATION

+5° +10° +15° +20° +25° +30° +35° +40° +45°

Betelgeuse
α

ω
ψ
φ λ
γ

ORION

GALACTIC

θ

β
6h

AURIGA

Capella
γ α

ι ζ ε
EQUATOR
5h

ι
τ
α Aldebaran
θ ε
γ υ

Hyades

μ
ASCENSION

4h
TO MAP 1

ν
λ TAURUS

χ
ξ
ε

δ ψ

η
Pleiades

o
PERSEUS

ν

Mirphak
α

RIGHT

ξ
o

β
Algol
3h

Menkar
α
λ

δ
ε
ARIES

ρ
π

θ

ANDROMEDA

2h
+50°

μ

ν
ξ
CETUS

γ

+45°

ECLIPTIC

+40°

+35°

Hamal
α

+30°

+5° +10° +15° +20° +25°

CLINATION

**The main
zodiacal**
constellation is
Taurus, with the
open clusters of

the Pleiades and
the Hyades. In the
southern
hemisphere, Lepus
(the Hare) can be
seen.

149

Star watching/5

Looking south in
the northern
hemisphere (north
in the southern),
the main
constellations
visible at midnight
on 23 January
(10 pm on 23
February and 8 pm
on 23 March) are
as shown.

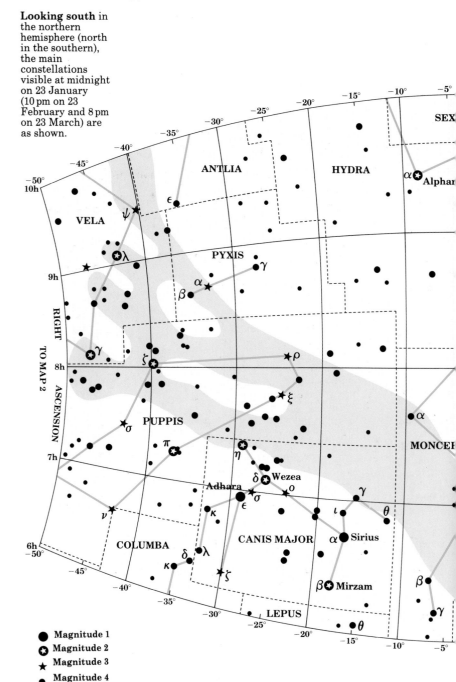

● Magnitude 1
✪ Magnitude 2
★ Magnitude 3
● Magnitude 4
• Magnitude 5

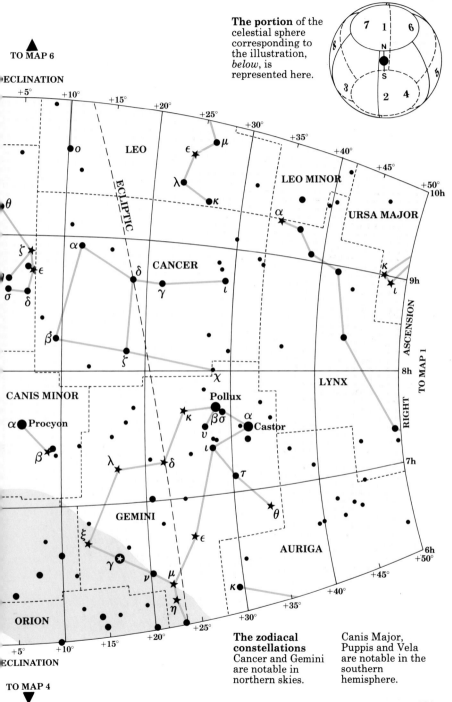

ECLINATION

+5° +10° +15° +20° +25° +30° +35° +40° +45° +50° 10h

The portion of the celestial sphere corresponding to the illustration, *below*, is represented here.

7 1 6
N
S
8
3 2 4 5

LEO

ο

ε μ

λ

κ

LEO MINOR

α

URSA MAJOR

θ

ζ
ε
α
σ
δ
δ

CANCER

δ

γ

ι

κ
ι 9h

β

ζ

χ 8h

CANIS MINOR

LYNX

Pollux

κ

β σ

α

α Procyon

Castor

υ

β

ι

τ 7h

λ

δ

θ

GEMINI

ε

AURIGA

ξ

γ

κ 6h
+50°

ν μ

+45°

η

+40°

+35°

+25° +30°

ORION

+5° +10° +15° +20°

ECLINATION

RIGHT ASCENSION

TO MAP 1

The zodiacal constellations
Cancer and Gemini are notable in northern skies.

Canis Major, Puppis and Vela are notable in the southern hemisphere.

Star watching/6

Looking south in the northern hemisphere (north in the southern), the main constellations visible at midnight on 24 March (10 pm on 23 April and 8 pm on 23 May) are as shown. The chief northern constellation is Canes Venatici, with the bright star Cor Caroli.

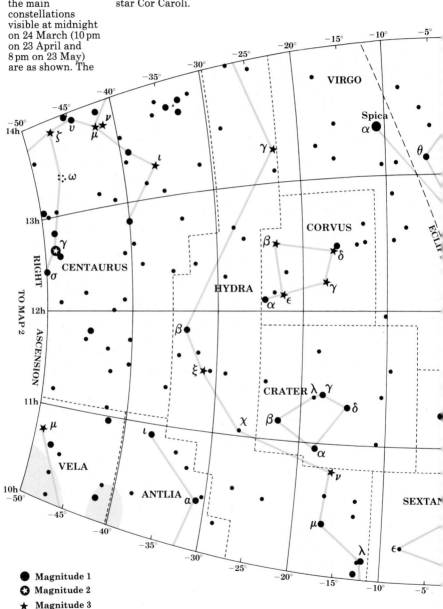

Magnitude 1
Magnitude 2
Magnitude 3
Magnitude 4
Magnitude 5

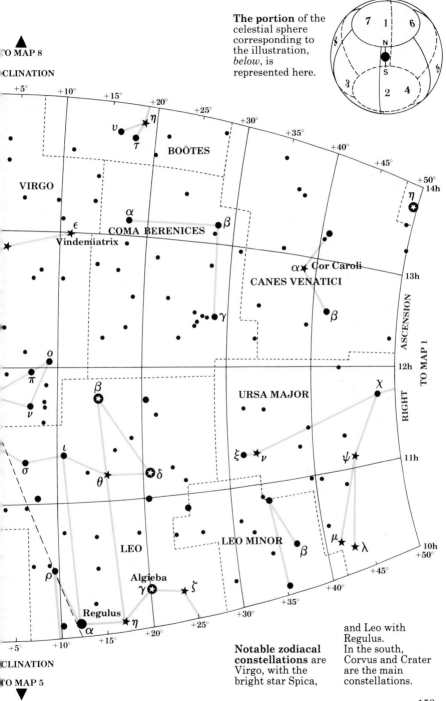

The portion of the
celestial sphere
corresponding to
the illustration,
below, is
represented here.

+5° +10° +15° +20° +25° +30° +35° +40° +45° +50°
14h

υ η
τ
BOÖTES

VIRGO

α
COMA BERENICES β

ε
Vindemiatrix

α Cor Caroli
CANES VENATICI
13h

γ

β

o

π
12h
β
URSA MAJOR
χ

ν

ι

σ

ξ ν
11h
ψ

θ δ

LEO MINOR

μ
β
λ
10h
+50°

LEO

ρ

+45°

Algieba
γ ζ
+40°

Regulus
η
α
+20° +25° +30° +35°

+5° +10° +15°

ASCENSION

TO MAP 1

RIGHT

Notable zodiacal
constellations are
Virgo, with the
bright star Spica,
and Leo with
Regulus.
In the south,
Corvus and Crater
are the main
constellations.

153

Star watching/7

Looking south in the northern hemisphere (north in the southern), the main constellations visible at midnight on 25 May (10 pm on 23 June and 8 pm on 23 July) are as shown. Notable constellations are Hercules, Boötes and the Crown, Corona Borealis.

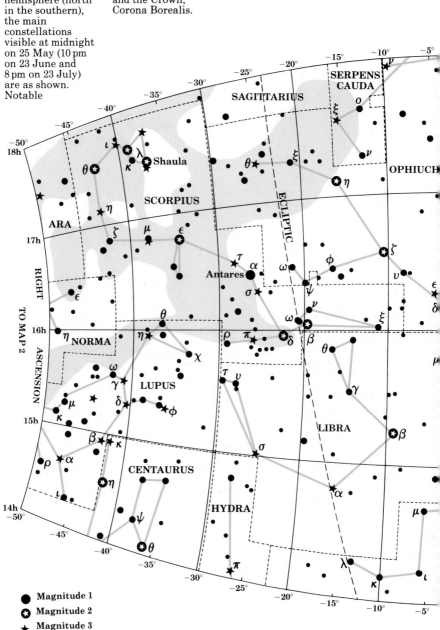

Magnitude 1
Magnitude 2
Magnitude 3
Magnitude 4
Magnitude 5

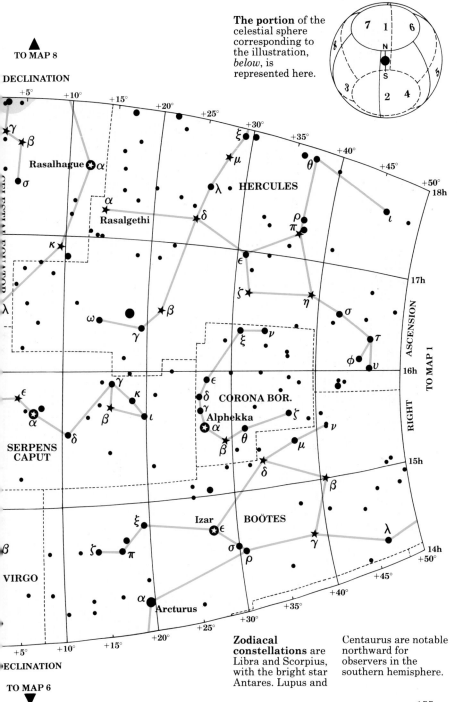

The portion of the celestial sphere corresponding to the illustration, *below*, is represented here.

7 1 6
8
N
S
3 2 4
5

+5° +10° +15° +20° +25° +30° +35° +40° +45° +50° 18h

γ
β
Rasalhague α
σ

α
Rasalgethi

κ
λ

μ
ξ
λ HERCULES
θ
ι

δ
ρ
π
ε

ζ η
σ
τ
φ
υ 16h

ω
γ β

ξ ν

ε

γ
ε
δ CORONA BOR.
γ
Alphekka
α
β θ
ζ
ν
μ
δ 15h
β

ASCENSION

TO MAP 1

RIGHT

17h

SERPENS CAPUT
ε
α
β ι
κ
γ
δ

ξ
Izar BOÖTES
λ
ζ π
ε
σ γ
ρ 14h
β +50°

VIRGO
α
Arcturus

+45°

+40°

+35°

+30°

+25°

+20°

+15°

+10°

+5°

ECLINATION

Zodiacal constellations are Libra and Scorpius, with the bright star Antares. Lupus and Centaurus are notable northward for observers in the southern hemisphere.

155

Star watching/8

Looking south in
the northern
hemisphere (north
in the southern),
the main
constellations
visible at midnight
on 25 July (10 pm
on 24 August and
8 pm on
23 September) are
as shown.

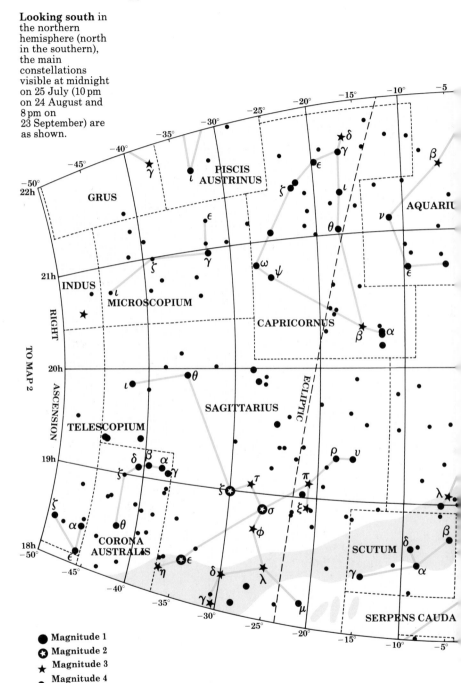

- ● Magnitude 1
- ✪ Magnitude 2
- ★ Magnitude 3
- • Magnitude 4
- · Magnitude 5

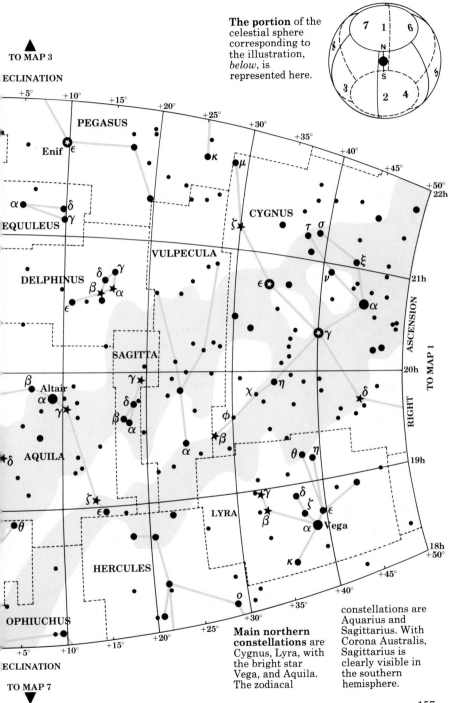

The **portion** of the
celestial sphere
corresponding to
the illustration,
below, is
represented here.

ECLINATION

+5°   +10°   +15°   +20°   +25°   +30°   +35°   +40°   +45°   +50°
22h

PEGASUS

Enif ε

κ   μ

CYGNUS

τ σ

α   δ
γ

ζ

ξ

EQUULEUS

VULPECULA

ν

21h

DELPHINUS   δ   γ

β   α

ε

α

γ

20h

SAGITTA

γ

η

χ

δ

β   Altair
α
γ

δ

β

φ

θ   η

19h

β

AQUILA

α

δ

ζ

ε

LYRA   γ

δ   ζ
β   ε

θ

α   Vega

18h
+50°

HERCULES

κ

+45°

o

+40°

OPHIUCHUS

+30°

+35°

+5°   +10°   +15°   +20°   +25°

ECLINATION

ASCENSION

TO MAP 1

RIGHT

**Main northern
constellations** are
Cygnus, Lyra, with
the bright star
Vega, and Aquila.
The zodiacal

constellations are
Aquarius and
Sagittarius. With
Corona Australis,
Sagittarius is
clearly visible in
the southern
hemisphere.

Guidance for observers

When you first turn your attention to the night sky, it is easy to think that the stars are simply tiny pin-points of light, randomly scattered in the inky blackness. But look through the star charts reproduced on pages 142–57 and you will soon realize that the stars can be recognized by the patterns they form and the way in which one group relates to another.

In the northern hemisphere, the important star guides are the Plough (Dipper) section of Ursa Major, the W of Cassiopeia, the T of Cepheus and the box of Perseus. Once these have been located, you should be able to find the Square of Pegasus, and from it Aries, Taurus and Cygnus.

Note, too, the constellations with bright stars – Lyra with Vega, Taurus with Aldebaran, Boötes with Arcturus, Auriga with Capella, Leo with Regulus and Canis Major with Sirius. Orion with Betelgeuse and Rigel is probably the easiest constellation to' find after the Plough (Dipper).

For observers in the southern hemisphere, Crux (the Southern Cross) acts as the first reference point. Nearby is Carina (the Keel) with Triangulum on the other side. Also notable are Aquarius, Vela (the Sails),

A planisphere shows on a flat surface the stars visible at certain dates and times.

The planisphere, *below left*, depicts the stars visible in the northern hemisphere at the latitude of London, 51½°N. The device is set for midnight in mid-December (10 pm in mid-January, 8 pm in mid-February). From New York it is possible to see further into the

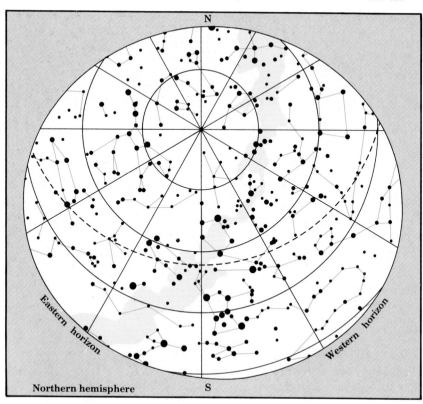

Northern hemisphere

the box, or 'body', of Corvus (the Raven) and the nearby box of Crater (the Cup).

Observers in the southern hemisphere also have the Large and Small Magellanic Clouds to guide them. The Small Cloud lies on the opposite side of the South Celestial Pole to Crux, and north of it are Tucana (the Toucan), then Phoenix and Cetus (the Whale). The Large Cloud is close to Volans, and north of it are Lepus (the Hare) and Orion.

Other guides in the night sky are the planets. When possible, look for Mars, Jupiter and Saturn.

To find out when specific constellations are visible, a planisphere is needed. Remember that at the same time one month later, each constellation will have moved westward by 30°. Stars at right ascension 0 hours will be due south at 10 pm on 22 March; those at RA 2 hours will be due south at the same time on 22 April, at RA 4 hours on 22 May, and so on. In other words, there is a 30° shift in right ascension every month. With this in mind, and by looking at the star charts, you can work out which constellations lie due south at any time of the year.

southern hemisphere. A planisphere for an observer at latitude 35°S –

Adelaide and Cape Town – *below right*, shows the constellations visible at midnight

in mid-June (10 pm in mid-July and 8 pm in mid-August). The many bright stars include Vega,

Achenar, Altair and the nearest of all stars, Rigil Kentaurus – α (alpha) Centauri.

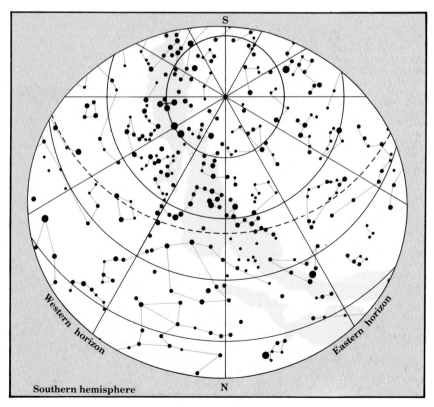

S

Western horizon

Eastern horizon

Southern hemisphere

N

Nebulae and clusters/1

If you are observant, you will be able to recognize many hazy patches of light in the night sky, which are known as nebulae (from the Latin *nebula*, meaning mist or cloud). The most famous of these is the Great Nebula in Orion, which is visible with the naked eye in Orion's sword around the star θ (theta) Orionis. In this area there is much bright gaseous material, including a number of 'emission' nebulae lit by hot, bright stars and 'reflection' nebulae, reflecting the light of bright stars near by, as well as dark, unlit gas and regions of interstellar dust.

Recent studies by the InfraRed Astronomical Satellite (IRAS) have made it clear that new stars are being born in some of the dark areas (as well as in other dark nebulous areas elsewhere near the plane of our Galaxy). The Orion region of the night sky is a magnificent sight in a telescope of almost any size and also in binoculars (see pp 192–7).

The Orion nebula is sometimes known simply as M42 (and a companion nebula near it as M43). These numbers refer to a catalogue of nebulae, galaxies and clusters compiled between 1771 and 1783 by a French

The double open cluster of stars centred around H and χ (chi) Persei can be seen, *right*. Their catalogue numbers are NGC 869 and 884 and they are less than 1° apart. They are associated and contain the same types of blue and red giant stars. The stars lie at a distance of some 7,100 light-years and are a magnificent sight in a 7.5-cm (3-in) telescope.

The Horsehead nebula in Orion is illustrated in the photograph, *opposite*. The 'horse's head' itself is a dark dust cloud, visible because it blots out light coming from the stars and glowing nebula behind it.

astronomer, Charles Messier. This work was prompted not because he had a particular interest in nebulae, but because he did not want to confuse such objects with comets (see pp 170–5), which also appear as hazy patches of light before they come close to the Sun and display their tails.

As telescopes improved, other catalogues besides Messier's were drawn up, two famous ones being the New General Catalogue (NGC) of 1888 and the Index Catalogues (IC) of 1895 and 1908. Many hazy objects have NGC as well as Messier numbers; thus M42 is also NGC1976. Together, these catalogues list more than 13,00(stial objects.

Even if you have only a modest telescope of about 15 cm (6 in), nebulae can be lovely sights. Besides M42 and M43 there is, for instance, M8 – the Lagoon nebula – in Sagittarius, which is another emission nebula together with a star cluster. As with M42, you can see M8 with the naked eye – it is a continuation of the curve formed by the stars ζ (zeta), ϕ (phi) and λ (lambda) Sagittarii – and in binoculars or a telescope it is beautiful. Another nebula and cluster is M16, close to γ (gamma) Scuti.

▷

Nebulae and clusters/2

A special class of nebulous objects is the planetary nebulae. They were given this name by William Herschel because they all looked to him like little round discs, greenish in colour and similar to the disc of Uranus. It is now known that they are nothing to do with planets, but are shells of shining gas blown off from a star; indeed every planetary nebula has a star at its centre. A few notable examples are M27, known as the Dumbbell nebula, just 3° north of γ (gamma) Sagittae, and M57, the Ring nebula, which lies about 7° southeast from the star α (alpha) Lyrae (Vega). Both are only visible in a telescope, and M27 needs a larger instrument than M57 – 15 cm (6 in) at least. In the southern hemisphere, there is the Helix nebula (NGC 7293) in Aquarius, which lies some 10° northeast of the first magnitude star Fomalhaut.

Clusters of stars are of two main kinds – open clusters and globular clusters. Open clusters are usually found in the vicinity of the spiral arms of the Galaxy and are made up of Population I stars. They are often beautiful to look at, as the Pleiades (M45) in the constellation of Taurus bear witness; these are visible to the

The Rosette nebula (NGC 2244), *right*, lies in the constellation Monoceros (the Unicorn) at a distance of some 4,500 light-years (1,380 parsecs). It is spherical and is illuminated by the energy emitted by bright but old stars (aged some 50,000 years) in its central regions.

The Trifid nebula (M20), *opposite*, is found in Sagittarius (the Archer). Lying some 2,300 light-years (700 parsecs) away, it appears divided by dark dust lanes into three main areas. It is composed of hydrogen lit by a bright, hot star embedded in its centre. Beside it is a blue nebula surrounding another hot star.

162

unaided eye and are sometimes referred to as the Seven Sisters.

If you have good eyesight, you will be able to distinguish the seven brightest of the Pleiades, but most people can see only six. Your view of them will be much improved by a pair of binoculars, but, in a telescope, beware of using too high a magnification because the cluster loses something of its effect. There are hundreds of stars in the cluster, and the brighter ones are of the hot, blue-white type, surrounded by glowing masses of gas. The cluster lies at a distance of some 400 light-years from Earth.

Almost as spectacular as the Pleiades are the nearby Hyades, though their brilliance to the unaided eye is somewhat reduced because they lie close to the bright, orange-coloured α (alpha) Tauri (Aldebaran). Moreover, Aldebaran's orange colour matches the colour of the brightest stars of the cluster. The Hyades cluster contains about 350 stars and is nearer to Earth than the Pleiades, being only 150 light-years away. It has no M or NGC number.

Two other clusters worthy of observation are, first, Praesepe in Cancer (M44), which is to be found midway

Nebulae and clusters/3

between α (alpha) Leonis (Regulus) and β (beta) Gemini (Pollux). It can be seen with the unaided eye, although you will need a telescope in order to see individual stars. The second is the Double Cluster (NGC869 and NGC884) in the sword handle of Perseus, in which the two components are only about 1° apart. It, too, is visible without a telescope, but with a small, 15-cm (6-in) telescope, you can make out quite clearly that the brighter stars are either red or blue giants – something well worth looking at. Observers situated in the southern hemisphere can see the lovely Beehive

cluster κ (kappa) Crucis, just south of β (beta) Crucis.

The second main type of cluster is the extremely compact globular star cluster. These formations are arranged in a spherical envelope around our Galaxy. They appear as hazy patches in binoculars or small 7.5-cm (3-in) telescopes, which is why Messier catalogued so many. In reality, they are vast, concentrated conglomerations, and a typical example may contain hundreds of thousands of yellow, orange and red stars. Although distances from Earth range between about 6,000 to well over 300,000 light-

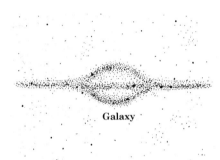

Galaxy

Our Galaxy, seen in a side view, *above,* shows the central bulge and thin spiral arm region. These are surrounded by a 'halo' of globular star clusters, which envelope the Galaxy in a kind of spherical coccoon. These clusters are all old and move in markedly elliptical orbits about the Galaxy's centre.

The globular star cluster M3 in Canes Venatici (the Hunting Dogs) is shown, *right.* This cluster has an apparent diameter of about one-fifth that of the Moon and an apparent magnitude of +6.4. A fine object even in a 7.5-cm (3-in) telescope, it contains tens of thousands of stars.

years (2 kpc to 100 kpc), a few are just visible to the unaided eye. But to see them at all clearly, a telescope must be used.

The most notable cluster for observers in the northern hemisphere is M13 in Hercules, 2.5° north of η (eta) Herculis. Its distance is 22,500 light-years and it contains approximately half a million stars. Three other globular clusters are well worth looking at – M5, M15 and M22 – though this by no means exhausts the possibilities. M5 can be found in Serpens Caput, some 6° southeast of α (alpha) Serpentis Caput. M15 is in Pegasus, some 4°

northeast of ε (epsilon) Pegasi, and M22 is in Sagittarius, close to a collection of four stars about 2° northeast of λ (lambda) Pegasi.

Of the 119 globular clusters so far identified in our Galaxy, the majority can be seen in the southern hemisphere, concentrated in the region of Sagittarius. Use the short list presented here as a guide only; a good star atlas will identify many more of observational interest. If you have only a small telescope, the open type of cluster tends to be a more rewarding sight, especially when it is superimposed on a nebula.

Double and variable stars/1

The proposition that stars could be made up of two components (double stars), was first put forward by the Englishman John Mitchell in 1767 and was proved correct by William Herschel in 1804. About half the stars visible in the sky are either doubles or multiple systems, with the component stars in orbit around each other. If you look at Mizar (see pp 142–57), the middle star of the handle of the Plough, or Dipper (Ursae Minoris), you may be able to see that it appears to be double. Look at it through binoculars or a telescope, and you will find that each star is itself a double star system.

Double stars are always worth looking at in a telescope, and this is especially so if the two components are of different colours. Though star colours are not usually particularly marked, the fact that both components are in view at the same time does bring out the differences between them remarkably well. An excellent example of this is β (beta) Cygni; here the brighter component (mag 3.1) is yellow and the dimmer (mag 5.1) is blue-white. Some other doubles to note are γ (gamma) Andromedae, whose components are orange (mag 2.18) and blue (mag 5.03); Castor, α

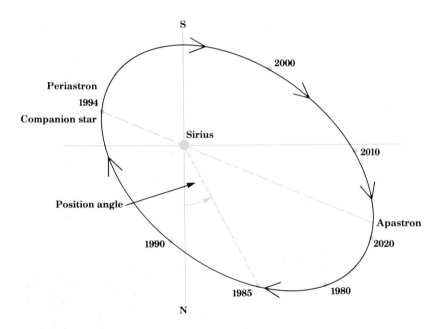

Observations of a binary star system are illustrated, *above*. In these observations, one of the stars, the 'primary', has been taken as stationary, and the relative movement of its companion

plotted by measuring its position angle and apparent separation from the primary. From the information obtained, the actual paths of both members of the binary, which

are really orbiting around each other, can be worked out. North is indicated at the base of the diagram because such observations are made through an astronomical telescope.

The dim but nearby binary, Kruger 60 in Cepheus, is shown, *opposite*. The intervals between pictures **1** and **2** show the binary after a gap of seven years, and between **2** and **3** after five.

(alpha) Gemini has components that are nearly the same in colour and brightness (mags 2.0 and 2.9), but it is worth looking at all the same. Each component is itself a double, although the components are so close together that they appear as single stars.

Such extremely close doubles (or, to be more precise, binary systems) can sometimes be detected by observing the cyclic shift in the lines of their spectra as one component comes toward Earth (giving a shift toward the blue end of the spectrum) and then moves away from Earth (redshift) as it orbits its companion star.

Sometimes, extremely close pairs pass material from one to the other. This may give rise to one star temporarily becoming particularly bright – what is termed a nova – or emitting X-rays, which can be detected only by an X-ray detecting satellite orbiting above the Earth. On rare occasions, a star that may or may not be a member of a binary system will explode; such a supernova may shine at 6 million times its previous brilliance.

In some binaries you can observe the components orbiting around each other. To do this, imagine the brighter component to be fixed and note the

▷

Name of star and constellation	Apparent magnitudes	Arc seconds between components	Right ascension (hrs min)		Decli- nation (° ')	
γ Andromedae	2.2, 5.0	10	02	04	+42	20
α Canum Venticorum	2.9, 5.4	20	12	56	+38	18
α¹, α² Capricorni	4.0, 3.7	376	20	18	−12	30
β Cygni	3.2, 5.4	35	19	31	+27	58
γ Delphini	4.2, 5.2	10	20	47	+16	07
ν Draconis	4.9, 4.9	62	17	32	+55	10
ψ Draconis	4.6, 5.8	31	17	42	+72	09
ε Lyrae	5.1, 5.4	208	18	44	+39	40
β Orionis	0.1, 0.7	3	05	15	−08	12
β Tucanae	4.4, 4.5	26	00	32	−62	58
ξ Ursae Majoris	2.3, 4.0	14	13	24	+54	55
α Ursae Minoris	2.0, 9.0	19	02	32	+89	16

Some double stars

1

2

3

Double and variable stars/2

position of the dimmer. As the months – or more likely the years – pass, you will notice a change. The easiest stars to observe doing this are wide doubles, but since these may take hundreds of years to complete one orbit, the change you see will be small. α (alpha) Gemini is such a star, and its orbital period is 350 years.

Optical doubles are pairs of stars that appear to be connected because they are in the same line of sight from Earth. There is, however, no other connection between them.

Some close binaries are situated so that their orbits appear edge-on; they

are known as eclipsing binaries. In such instances, the star seems to vary in brightness because for part of the time light is received from both components; at other times one component eclipses the other, and light from only one is received. Such a star is β (beta) Persei (Algol), also known as the Demon Star; another is β (beta) Lyrae. In the southern hemisphere there is ζ (zeta) Phoenicis, which varies between mag 3.6 and 4.1 every 1.67 days.

Widely spaced binaries are a better subject if you have only a small telescope. For these you may need a fairly high magnification, but for

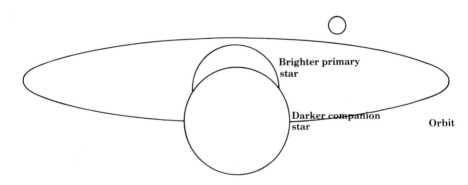

An eclipsing binary, *above,* can appear to an observer as a variable star. It is recognizable as a binary because of

the particular way in which its light varies. When the dimmer, orbiting companion star eclipses the bright primary, the

binary shows a sudden, deep reduction in light. It then rises swiftly to its normal brightness because light from both

components reaches Earth. A 'secondary minimum' occurs when the primary eclipses the companion.

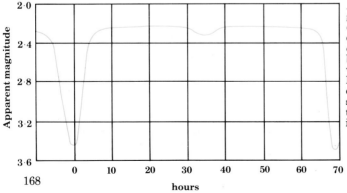

A characteristic graph, or light curve, is used for depicting the changing brightness of a variable star. The light curve of an eclipsing binary, such as the one in the drawing, *above,* is shown, *left.*

close binaries, an even higher magnification may be necessary. An instrument of 21.5 cm (8½ in) is about the smallest size practicable.

Not all stars that vary in brightness are eclipsing binaries; most are true variables. It is possible that every star becomes a variable at some point as it evolves. In some, the amount of energy radiated from their surfaces changes continually as a result of their expanding and contracting slightly. Some vary their light output frequently. Such stars include the Cepheid variables, named after δ (delta) Cephei (mag range from 3.5 to 4.4), which has a period of variation of 5 days, 8 hours, 53 minutes. You can observe the change even with the unaided eye. Cepheids are important because their period of variation depends on their true brightness – the brighter they really are, the more slowly their brightness changes.

A different type of variable is typified by the red star o (omicron) Ceti (Mira). When bright (mag 3.0), it is clearly visible to the unaided eye, but when dim (mag 8.5 or even 9.5), it can be seen only with binoculars or a telescope. Mira's fluctuation period is approximately 330 days.

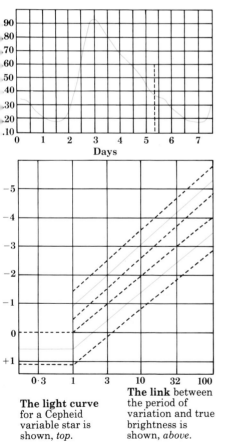

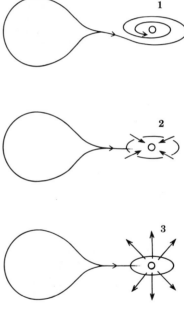

The light curve for a Cepheid variable star is shown, *top*.

The link between the period of variation and true brightness is shown, *above*.

The transfer of material in a close binary system, in which the companion star is a white dwarf, is shown, *above*. 1 Material is transferred from the more massive star to the smaller one. 2 The material is heated by the smaller star's white-hot surface. 3 The material flares to immense brightness, known as a nova, for a short time until it is burned away.

169

Shooting stars and comets/1

The first thing to note about shooting stars is that they are not stars at all; they are relatively small pieces of stony material from space which have been captured by the gravitational pull of the Earth. The visible trace left in the sky by a shooting star is the result of this material rushing through the Earth's atmosphere.

On this final part of its journey, it is heated by friction with the air, and its outer atoms vaporize. These atoms collide with those of the surrounding air, making them glow; it is the glowing air trail that is seen as the path of a shooting star or, more accurately, meteor. The velocity of this meteoric material is so high – anything between 10 to 70 km/sec (7 to 45 mls/sec) – that in most instances the meteor is burned away before it even reaches ground level. Usually it is completely vaporized some 80 to 120 km (50 to 75 mls) above the ground, and the most that gets through is fine dust. Most meteor trails are dim, but sometimes the material may be massive enough to generate a really bright trail; it is then known as a fireball.

On rare occasions, the lump of material will be too large to be consumed completely during its fall. Some rocky material will then land, but it is often difficult to recover the pieces of what has now become known as a meteorite. Some such pieces have, however, been found and can be seen on display in many museums.

A meteorite can cause damage, and in Arizona in the USA there is a vast crater caused by such a fall. The crater's diameter is 1.2 km (4,000 ft) and it is about 180 m (600 ft) deep inside its rim, which rises some 60 m (200 ft) above the surrounding desert landscape. The damage caused here would have extended for many kilometres around the area of the crater due to the blast from the hot gases accompanying the rocky lump.

Meteorites are divided into three

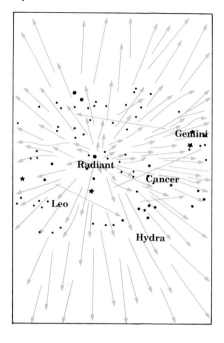

The radiant point of the Leonid meteors can be seen in the chart, *above*. The meteor stream appears to issue from a point in the constellation of Leo, hence the name of the shower. The meteors are, however, nowhere near that constellation but are moving through the Earth's atmosphere.

A meteor radiant is the result of an optical illusion. In the photograph, *opposite*, the parallel edges of the road seem to meet and disappear into the starry distance. The point where parallel lines appear to meet is known as the vanishing point. The same type of effect produces the meteor radiant.

▷

Shooting stars and comets/2

main types – stony, stony and iron and iron. The stony type is most common, making up approximately 60 per cent of the total; stony-iron types make up only about 10 per cent, and 30 per cent are iron types. The stony meteorite, as well as being most common, is also the most brittle, and many break into small fragments in the atmosphere.

Meteorites can be seen on any clear night, but mostly they appear merely as faint streaks of light that last for a fraction of a second. On average there will be about 6 meteors per hour early in the evening, increasing to something like 14 per hour in the early

hours before dawn. The reason for this difference is that the Earth is orbiting around the Sun and rotating at the same time. Thus, early in the evening, we are trailing behind meteors coming in from the night side of the Earth and can see only the quick ones; as the night progresses, we meet them increasingly head-on and can detect those with less speed as well.

Not all meteors, however, are of this chance, or sporadic, nature; some are to be found in streams, orbiting the Sun. These appear at various times during the year, when the orbits of the Earth and meteor stream coincide.

Path of Halley's Comet around the Solar System

Neptune's orbit

Feb 1986

Sun

1987

Earth's orbit

1985

201

The Bayeux Tapestry illustrating the Norman Conquest of Britain in 1066 depicts Halley's comet. King Harold is being warned about its appearance, which was thought to presage disaster.

The most famous of these streams is the Leonids, which arrive in November, but there are many others. On such an occasion, the hourly rate of detectable meteors will rise from the normal 6 to more than 40 per hour, sometimes even exceeding 100. The meteors in such a shower always appear to radiate outward in all directions from a fixed point or radiant. This is purely an optical effect, since the meteors are, in fact, travelling in parallel lines. The illusion is similar to that created when you look at a road; the edges of it appear to spread out from some distant point.

A meteor shower is named after the constellation in which its radiant lies. Thus the Leonids have their radiant in Leo, the Geminids in Gemini, the Perseids in Perseus, and so on.

If you have the chance to observe meteors, note the points in the sky (referred to nearby stars) at which the trail begins and ends. Note, too, the time and the brightness by comparison with nearby stars.

Most meteor showers have been found to follow the orbits of the comets. This is not surprising because a comet is a compact collection of rocky material (the nucleus) and

▷

Orbit of Halley's Comet

The Great Comet of 1843, *above*, was a 'sun-grazer', which came very close to the Sun during its perihelion. As a result, it grew an immensely long, straight tail. Comets always arouse popular interest; public imagination often harks back to the old idea that comets are associated with the souls of famous men, as the cartoon, *right*, from *Punch*, May 1910, illustrates.

2024

Halley's Comet has an orbit whose period is 76 years. It comes close to the Sun during perihelion but goes out beyond Neptune's orbit at aphelion.

Shooting stars and comets/3

frozen gases orbiting the Sun, usually in an elliptical path steeply inclined to the plane of the Solar System. At a distance of about two a.u. from the Sun, ice within the comet begins to vaporize and form a cloud, known as a coma, that envelops the nucleus.

As the comet approaches closer to the Sun (nears perihelion), the frozen gases are heated and it is often seen to have a tail. This always points away from the Sun, irrespective of the direction in which the comet is moving. Most comets are extremely faint, do not have a tail and are visible only in a telescope. Unlike a meteor, a comet remains on view night after night, sometimes for months. When one can be seen with the unaided eye, it is a magnificent sight.

The orbital periods of comets vary widely, from a few years for a short-period comet to 200 years or much more for a long-period meteor. The most famous is Halley's Comet, which has an orbital period of some 76 years and is due to appear next in the winter of 1985. At that time, it will be visible in northern skies in Taurus and near the Square of Pegasus before it becomes hidden by the Sun. It will reappear in the skies of the southern

Visibility of Halley's Comet 1985/6

6 Nov Pleiades 26 Nov Square of Pegasus
Aldebaran Taurus Altair Aquila
16 Dec
Orion Cetus 25 Jan Perihelion of comet
Aquarius 12 Feb
Position of Sun at perihelion of comet 6 March Sagittarius
Capricornus 21 March Antares
For observers in the northern hemisphere 1 Ap

hemisphere in the spring of 1986. At its last appearance, in 1910, it was a beautiful sight; but this time it is expected to be less spectacular, for, at the angle at which it will be seen, its tail will appear to be foreshortened.

Spacecraft will be launched to approach the comet to determine whether current theories of the nature of comets are correct. Certainly it is now known that, by cosmic standards, comets are ephemeral bodies. On each orbit a little of their mass is lost as it is heated by the Sun until, eventually, the nucleus breaks up.

Comets are now named after their discoverer. Halley's, however, was first definitely recorded in 240 BC. It bears Halley's name only because he discovered that it was periodic – the first comet to be confirmed as a regular visitor. Comets are also given a year and a letter indicating their order of discovery during that year, and a year and a Roman numeral indicating the order in which they reach perihelion. Thus Comet Kohoutek was discovered by Lubos Kohoutek; it was given designation 1973f, being the sixth comet discovered that year. Later it became 1974XII, being the twelfth comet to reach perihelion in 1974.

The path of Halley's Comet as it will appear in 1985/6 is illustrated, *left,* threading its way through the constellations. The comet will be best seen by observers in the southern hemisphere early in 1986, but is unlikely to be as bright as it was at its last appearance in 1910. In the photograph, *below,* the comet is shown when it appeared near Venus and its brightness can be gauged by comparison with that brilliant object. The photograph is presented on its side so that the comet's tail can be included.

Artificial satellites/1

Since October 1957, when the USSR launched Sputnik 1, and the following January, when Explorer 1 was put into orbit by the United States, artificial satellites have been circling the Earth. Although most of the hundred or more operational satellites at present in orbit are for military use, there are many dedicated to peaceful purposes.

Some of the most interesting research satellites include those which are designed to assist in prediction of the weather and to give warning of approaching fronts, depressions, cyclones and so on. Many televised weather reports, worldwide, now make use of these satellite pictures to help explain the most likely short- and medium-term weather patterns. Other types of satellite scan the Earth's surface, pinpointing promising sites of mineral deposits or giving early warning of developing crop diseases.

Astronomical satellites, too, are examining the universe in the infrared and at X-ray and gamma-ray wavelengths. These satellites produce information unobtainable in any other way. The Earth's atmosphere filters out most or all of such wavelengths emitted by celestial bodies, which

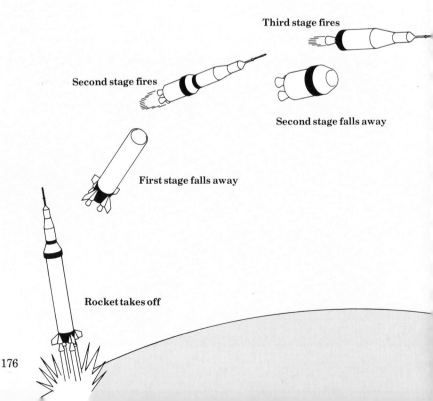

Third stage fires

Second stage fires

Second stage falls away

First stage falls away

Rocket takes off

means that ground-based observers cannot detect them.

The geostationary communications satellites are of special interest because they have orbits which are at such a height that they orbit the Earth once every 24 hours. The result is that they appear to remain stationary above fixed points on the Earth's surface, making them ideal vehicles for efficient long-distance communication.

If you know when to look, and sometimes just by chance, you can see such satellites, usually a little after sunset or just before sunrise. Being so high above the Earth's surface, they reflect back the Sun's light, even though the Sun is below the horizon.

The detailed study of the behaviour of satellites in orbit can be useful. By timing the motion of satellites across the background of fixed stars, vital information can be obtained about the precise shape of the Earth; it also permits more accurate predictions of satellite appearances to be made.

The basic method of observation is to use predictions so that you can be looking in the correct direction near to the right time. A stop-watch with an accuracy of at least 0.5 seconds is essential. Set the watch by radio time

▷

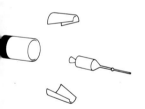

Capsule separates and enters orbit

A rocket must attain a velocity of more than 11 km/sec (7 mls/sec) before it can escape the gravitational pull of the Earth. In a typical launch of a three-stage rocket, the entire heavy rocket is lifted off the ground by the first (bottom) stage. This drops away as soon as it has consumed its liquid fuel. The rocket continues on its way, now propelled by the second stage. It, too, drops away once its fuel is exhausted. The third and final stage now fires and puts the 'payload' of instruments or astronauts into a stable orbit before detaching and falling away.

Artificial satellites/2

signals or by the telephone system's speaking clock. The latter is the more satisfactory method, since the stopwatch can be set just before observation begins.

Gauge the position of a satellite by imagining a line drawn between two fairly close stars between which the satellite passes, or by imagining a vertical line drawn down from a particular star. In either instance, it is the precise moment at which the satellite crosses this line that is important. The stars can be identified from a star atlas, and it is a good idea to draw a sketch map of the star field chosen and to indicate the satellite's track on it.

Observers find that identification of the position is not difficult; the great problem is the accurate timing of the transit of the satellite. As in all practical astronomical work, this is an example of the need for practice and constant repetition in order to produce reliable results.

Artificial satellites vary in brightness, depending on the height at which they are orbiting and on their size. They can best be divided into three classes: bright satellites visible to the unaided eye – usually brighter than mag 3; those that are dimmer, with magnitudes between 3 and 9; and extremely faint satellites. With the unaided eye it is possible to see down to magnitude 6, although below 3 or 4, you will need to know where to look. Satellites dimmer than magnitude 6 require binoculars or a telescope if they are to be seen and, in any event, it is important to know where and when the satellite is expected to appear from tables published in some journals and daily newspapers.

All satellites transmit their information back to Earth by radio signals, some of which have been picked up by amateur astronomers. Generally speaking, however, a powerful shortwave receiver and a large aerial system, or antenna, are required.

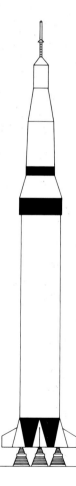

In a three-stage Saturn rocket, *left,* the upper part contains the Apollo command, service and lunar modules. At the very top is the launch escape system.

The second stage of the Saturn lies just below the much smaller third stage. The latter contains one jet only, but the second stage has five jets through which its liquid fuel can fire.

The first stage is the largest. This, too, has five jets through which its fuel of kerosene and liquid oxygen fires. The photograph, *opposite,* shows the launch of a Centaur rocket.

179

OBSERVING & THE SKY

The naked eye

Initially, the most useful instrument for observing the sky by day or by night is the naked eye. For the novice observer, it is important to become familiar with the general aspects of the sky, and the wide field of view provided by the human eyes is ideal for this. Once this familiarity is established, binoculars (see pp 192–5) or a telescope (see pp 196–201) are indispensable for detailed observations.

The first thing to notice when observing the night sky is that the stars all remain in the same patterns or constellations (see pp 142–57). These rise in the east and set in the west but, because the Earth is moving around the Sun, they do so four minutes earlier each night. The result is that over the year different constellations can be seen in the night sky. If, for example, Orion were due south at midnight early in December, then early in April, Virgo would be in this position.

The Moon does not behave quite like the stars. Because of its motion as it orbits the Earth once every month, the Moon rises some 50 minutes later each night. The planets alter their positions among the stars, too, but they do this more slowly. They thus appear to rise approximately four minutes earlier each night.

Lunar eclipses, which occur two or three times every year (see pp 204–15 for dates), are easy to observe with the naked eye. These happen when the Moon is full and moves into the Earth's shadow. Other readily observable events include meteors, especially at the time of a meteor shower, and a number of variable stars (see pp 166–9).

At those times half-way between day and night, at sunset and dawn, the planets Venus and Mercury can be seen, also Saturn and Jupiter and the reddish planet Mars. Solar eclipses are also regular events, but when observing the Sun take particular care so as not to damage your eyes.

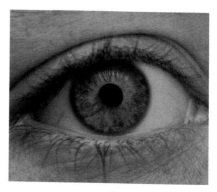

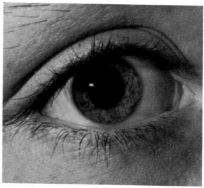

The human eye is a marvellously flexible instrument. In the middle of the eye is the iris with the pupil at the centre, through which light enters. The amount of light is controlled by contraction, *top*, or dilation, *above*, of the pupil by the muscles of the iris. An elastic, transparent disc just behind the pupil is the lens, which brings light into focus on the retina at the rear of the eye.

To the naked eye, clouds look solid, *opposite*, only because they are distant. A cloud is really composed of millions of water droplets, which reflect and refract light as it passes through them. Dense clouds will look almost black to an observer on the ground because light cannot penetrate. Viewed from above, by reflected light, they may have a much brighter appearance – this is their 'silver lining'.

Viewing the Sun safely

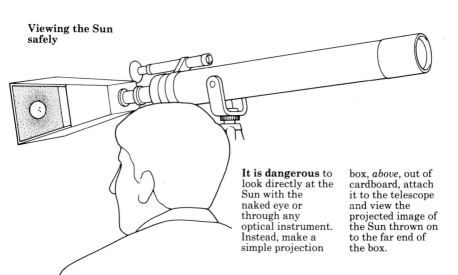

It is dangerous to look directly at the Sun with the naked eye or through any optical instrument. Instead, make a simple projection box, *above,* out of cardboard, attach it to the telescope and view the projected image of the Sun thrown on to the far end of the box.

Photographing the daylight sky/1

Successful photographs of the daylight sky depend, to a large degree, on an understanding of the nature of light itself. Although usually described as white, light is a mixture of different wavelengths, which, when separated, perhaps by water droplets in the atmosphere, can be seen as the individual colours of a rainbow (see pp 14–19). Other factors, such as the time of day and the time of year, also affect the appearance of light and thus the objects illuminated.

Practically any camera is suitable for daylight photography. Ideally, it should be able to accept a range of lenses from a wide-angle lens of 24 mm for broad sky panoramas to a telephoto lens of about 200 mm for picking out distant cloud formations.

Filters that are commonly used include a colourless UV (ultraviolet) filter, designed to minimize the effects of atmospheric haze; a polarizing filter to remove contrast-reducing polarized light from a scene; and a range of yellow, red and green filters of different strengths to increase the contrast between clouds and sky. Unless distorted colours are wanted, coloured filters should be used only with black and white film.

The choice between colour and black and white film is largely a matter of personal preference. Subjects such as rainbows and coloured skies obviously benefit from colour film, but studies of clouds and the effects of wind, rain, frost and snow, for example, can equally well be recorded in black and white. At sunrise and sunset, when the red content of visible light is high, a panchromatic black and white film should be used. Orthochromatic film is insensitive to deep orange and red light. A slow, ISO 25–80 film is usually adequate for sky photographs.

Pictures of the sky are often disappointing, lacking in contrast and featureless. Contrast is dependent upon

Low, threatening clouds form the perfect backdrop to the ancient stones of Castlerigg, near Keswick in northern England. The photograph was taken in the late afternoon. With the Sun low in the sky, strong, contrasting areas of light and shade were created, which were further strengthened by an orange filter over a 28-mm wide-angle lens. Colour filters should be used only with black and white, in this case an ISO 32 black and white transparency film.

▷

Filter recommendations

This table contains recommended filter colours to correct or increase contrast between clouds and sky. These filter recommendations apply only to panchromatic black and white film.

Subject	Result	
	Natural	Exaggerated
Bright hazy sky	Medium yellow/ green	Red
Pale sky, low contrast	Light yellow/ green	Red
Blue sky, good contrast	Light yellow	Orange
Dark sky, strong contrast	No filter	Medium yellow/ green
Sunset, red sky	No filter	Blue-green
Sunset, blue sky	Light yellow	Red

Photographing the daylight sky/2

the type of cloud, the direction of the light and the exposure set on the camera. When taking a light reading, take care to exclude the Sun, especially with a through-the-lens metering system. Remember that looking at the Sun, even on the camera focusing screen, can damage your eyes. A partially obscured Sun can also be dangerous and will indicate on the meter an exposure value that would render most of the scene underexposed.

One of the easiest cloud types to photograph is the bright cumulus, occurring as it often does in a dark blue sky. More difficult are the wispy cirrus clouds, which are usually found in a pale sky. The low-level nimbus often present most problems. These grey clouds lack brilliance and modelling and reproduce as though they were flat. Toward the horizon, the sky appears lighter and successful cloud photography becomes harder.

Atmospheric haze can sometimes be used to good effect as an indicator of depth and perspective in pictures, but, more often, it merely obscures the true subject. There are, however, times of day when haze can largely be avoided. Haze tends to increase as the Sun warms the atmosphere, so shoot in the first few hours after sunrise and toward the end of the day.

Many dramatic photographs depend on being in the right place at the right time. Lightning, for example, is not difficult to record, but a tripod and a camera with a T-setting or B-setting are required. A lightning flash produces enough light to allow a slow to medium (ISO 80–200) film and an aperture of about f 11. As a daylight technique, however, this is effective only toward dusk or early in the morning. At these times it is dark enough to permit an exposure of a few seconds. If the lightning is overhead, make sure that you, or your camera and tripod, are not the highest objects in the immediate vicinity.

Black and white film can turn a simple descriptive photograph into a more powerful image, *right.* Shape, form and textures become more important when colour is removed from the scene. Exposure was for 1/125 sec using a 28 mm wide-angle lens on a 35 mm camera.

At midday, with the Sun high in the sky, *right,* daylight is at its strongest. Only short shadows are cast and facial detail is lost in the top lighting. Light from the Sun has its shortest journey through the atmosphere at this time and tonal contrasts between clouds and sky can be intense.

Photographing the night sky/1

Photographs of objects in the night sky are more complicated to achieve than those taken in daylight (see pp 184–7). But except for the addition of a sturdy tripod, all that you need to get started is a camera, a lens and either colour or black and white film.

Generally, light levels at night are extremely low, and a tripod is essential to hold the camera completely still during the long exposure times that are often required. A shutter release cable to trigger the shutter is an inexpensive extra item and makes any camera movement much less likely.

The best all-round camera for amateur astronomical photography is the single lens reflex (SLR). With this camera, it is possible to change from a normal lens for general pictures of starry skies to a telephoto lens for reasonably detailed studies of the lunar landscape.

Many SLRs feature automatic light measurement and exposure, but this facility is not designed to cope with night-time levels of illumination. Use either a manual model, or one that can be switched to manual operation, and select the T-setting on the shutter control ring. Once the shutter is triggered, it will remain open until the shutter release is pressed again.

There are two techniques for taking pictures of the night sky. The first involves using a stationary camera, the second a driven camera.

Using the stationary technique, place the camera on the tripod and set the camera to focus on infinity. Next, select an area of sky and then make the exposure. Due to the rotation of the Earth, however, an exposure longer than about 30 seconds will record stars as streaks rather than pinpoints of light. If you include an area of sky containing the celestial poles and expose for about five minutes the resulting photograph will show circular star trails.

The stationary camera technique

The Full Moon, *above,* photographed over the Parthenon, Athens, using a 400 mm lens and a 2-sec exposure. The Full Moon is not the best time for detailed lunar observations because the level of illumination is too high. Some detail has also been lost due to halation – light reflecting back from the base of the film.

A telescope adaptor ring, *above,* allows any single lens reflex camera to form part of a sophisticated recording system of celestial objects. Use the chart, *opposite,* as a guide to exposure times.

Comet Burnham 1960 II, *opposite,* photographed on 27 April 1959. An exposure of 20 sec was used and a 15-cm (6-in) telescope.

▷

Exposure recommendations

This table is an approximate guide only. Ambient light levels, the brightness of the sky background, general atmospheric conditions, as well as differences in film processing, all contribute to make exposure times extremely variable. Whenever possible, it is advisable to bracket exposures. This involves making a series of exposures on either side of the one determined to be correct.

Object	Mount	Film speed	Aperture	Exposure time
Star trails/ comets	Tripod	ISO 64	f2–f4	Up to 30 min
Meteors	Tripod	ISO 64–200	f5.6	10–30 min
Full Moon	Optional	ISO 64	f8	1/250–1/500 sec
Quarter Moon	Optional	ISO 64	f5.6	1/125–1/250 sec
Lunar eclipse:				
Half shadow	Tripod	ISO 200	f4	1 sec
Near totality	Tripod	ISO 200	f2.8	2 sec
Artificial satellites	Tripod	ISO 200	f4	10–30 min
Stars/comets	Driven	ISO 200	f5.6	1 min–1 hr
Star clusters/ nebulae/ galaxies	Driven	ISO 200–400	f5.6–f8	10 min–1 hr

Photographing the night sky/2

can also be used to record the passage of bright comets, meteors and artificial satellites. For these objects, a fast film should be used and an exposure time of between 5 and 30 minutes with the camera lens fully open. Again, due to the rotation of the Earth, the photographic image will not be sharp.

To obtain sharp images of celestial bodies, a driven camera is essential. This means attaching the camera to a special mount with a motorized drive system, such as is used with telescopes (see pp 196–9). In this way, the mount moves to compensate for the rotation of the Earth, thus keeping the object under study in the same place relative to the camera.

For detailed recordings of the Moon, planets, star clusters, nebulae and galaxies, the camera lens has to be supplemented or replaced by a telescope. The simplest method is to position the eyepiece of a telescope close to the camera lens. With this system, it is difficult to check focus, especially if the two instruments are not precisely aligned. Also, the additional glass surfaces introduced into the light path cause a degree of scattering, with a resulting loss of image sharpness.

A better method is the eyepiece-projection system, in which the camera lens is removed and the camera body attached (via a coupling device) to the telescope eyepiece. This removes one set of lens surfaces, with a resulting gain in image sharpness.

For maximum image contrast, the prime-focus method should be considered. In this, the camera is attached to the telescope from which the eyepiece has been removed, and the object lens of the telescope (see pp 196–7) focuses the light directly on the film plane. Because it is the eyepiece that performs the enlarging function, the resulting image is small, but extremely bright and sharp. This makes it ideal for enlarging as a print or projecting as a slide.

The progress of a total eclipse is charted in a sequence of $\frac{1}{4}$-sec exposures of the Moon, *above*. The photographs were taken at approximately 5-min intervals on 12/13 April 1949.

A total lunar eclipse, *right,* recorded on 5 July 1982 as a continuous 4-hr exposure.

An equatorial drive added to a telescope and camera set-up will allow you to track the movement of celestial bodies during long exposures.

Binoculars/1

For any astronomer, whether amateur or professional, binoculars are standard items of equipment. Binoculars form a valuable link between the naked eye and a telescope, and examples of almost every type of celestial object are visible with them.

Whether you are sweeping the Milky Way, scanning star fields or observing the passage of a bright comet, a good pair of binoculars will present a magnificent, wide-field view of the night sky. And because both eyes are being used, the images you observe have a real appearance of depth. Observing in this way is a great deal more restful than using a telescope (see pp 196–7), and thus longer observation periods are possible.

Significant astronomical discoveries have been made by knowledgeable amateur observers using very powerful binoculars; for example, many comets have been found. Variable star observers also find binoculars useful, since many variables (see pp 166–9), are well within range.

Binoculars with magnifications of more than 7 times should be mounted. The reason for this is that the more powerful types are heavy and difficult to hold steady, and even the pulsing of one's blood can cause the binoculars to waver. The most common method of mounting binoculars is to use a camera tripod. Adaptors are available that clamp around the central axle of the binocular and attach to the tripod camera mounting bolt.

These are two types of binocular – roof prism and Porro prism. The roof prism type is usually smaller and easier to handle because its direct-light-path design allows a more compact construction. Most astronomers, however, are more familiar with the Porro prism variety. These instruments have the typical binocular shape but contain bulkier prisms. These are present in order to orientate correctly the image (which would

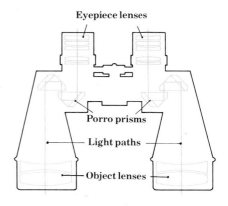

Eyepiece lenses
Porro prisms
Light paths
Object lenses

Porro prism binoculars use two pairs of prisms to produce an image that corresponds to normal vision. As a result, they are suitable for both celestial and terrestrial use, but are bulky and can be awkward to hold for long periods of observation.

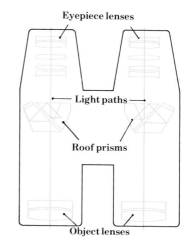

Eyepiece lenses
Light paths
Roof prisms
Object lenses

Roof prism binoculars are optically simpler than Porro prism types. The single prisms produce an inverted image, making them suitable only for celestial use. They are relatively light in weight and suitable for long periods of observation.

▷

Binocular types

Binoculars are classed according to magnification and aperture. A size of 7 × 50, the most suitable size for general astronomical work, indicates a magnifying power of 7 times and an object glass diameter of 50 mm. The exit pupil is the bright disc of light emitted by the eyepiece, a measurement obtained by dividing the aperture by the magnification. For best results, the size of the exit pupil should match that of a fully dilated human pupil (about 7 mm).

Type	Magnification	Aperture (mm)	Exit pupil (mm)
6 × 30	6	30	5
7 × 35	7	35	5
7 × 50	7	50	7
10 × 50	10	50	5
10 × 70	10	70	7
11 × 80	11	80	7.5
20 × 80	20	80	4

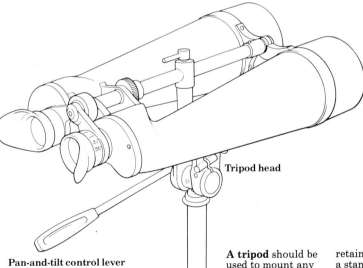

Tripod head

Pan-and-tilt control lever

Tripod

A tripod should be used to mount any binoculars with a magnification of more than about 7 times. Some large instruments may have a threaded hole designed to accept the retaining screw of a standard photographic tripod. If not, adaptors are available that clamp on to the central axle of the binoculars.

Binoculars/2

otherwise appear inverted). Porro prism binoculars are thus suitable for both terrestrial and celestial use.

In good-quality binoculars, all the optical surfaces, including the prisms, will be coated to improve light transmission and image brightness. This feature can make the difference between searching for a faint nebula and actually seeing it.

Before making a purchase, examine the binoculars carefully. Once you have decided on the appropriate magnification, compare prices of the same model in different stores. Make sure that you handle them and that the weight and bulk are within your limits of tolerance. Also ensure that the mechanical quality is up to standard; the central axle should be firm and the focusing mechanism smooth. The best way to test the optics is to look at a few distant objects. The image should be crisp all the way to the edges of the field of view. Watch also for colour rings around bright object.

Colour rings around the image indicate chromatic aberration, in which the component colours of the light source are not being brought into common focus. This fault is standard with a refracting instrument, including refracting telescopes, but it should not approach a noticeable level. To overcome the worst effects of chromatic aberration, multi-element lenses are used. These lenses are composed of a variety of glasses, each with a different refractive index, or ability to bend light.

Once purchased, all binoculars should be treated with care. Provided that the external glass surfaces are covered after use by the caps supplied, and that the binoculars are not subjected to any sharp knocks (which could loosen the lens elements) they should last a lifetime. If the lens surfaces are coated, avoid using any type of liquid lens-cleaning fluid; remove loose dust, using a blower brush.

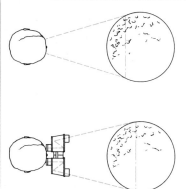

Stereoscopic vision
Observing with both eyes through binoculars creates more of a sense of depth than through a telescope because a little of the sides of the object under scrutiny become visible. The widely spaced object lenses of binoculars reinforce this effect.

Coated glass surfaces
Better-quality optical instruments have all their internal and external air-to-glass surfaces coated with a substance designed to improve light transmission. This should eliminate nearly all glare and reflections and increase the brightness of subjects such as the Moon, *above*.

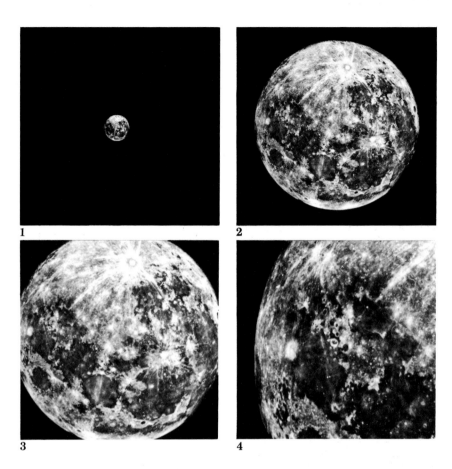

1

2

3

4

Choosing binoculars is not a matter of magnification alone, but also of the correct magnification for the object to be observed. The larger the magnification, the less the field of view of the instrument, as can be seen, *above*. The Moon as it could appear with the naked eye, **1** shows a complete disc. With 7 × 35 binoculars, **2** the disc is much larger, and since this size increases with 9 × 35 binoculars, **3** it can be seen that the field of view becomes much narrower. With a 20 × 50 instrument, **4** the complete disc is no longer visible.

Binoculars for use at night are known as night glasses and have larger objective lenses than their daytime equivalents. It is the object glass that collects the available light.

Optical alignment is important in binoculars. Unless both barrels and all the internal optics are perfectly parallel, the two images will not merge into one circle. Misaligned optics will have a detrimental effect on observable detail and can cause eye strain.

Telescopes

There are essentially two types of telescope – refractors and reflectors. In a refracting system, a lens is used to collect the light and bend it to a point of focus, while a reflector makes use of a specially curved mirror to collect and reflect the light back to the focal point. In each system, an eyepiece is placed at the focal point to magnify the image produced. There are advantages and disadvantages with both types, and before either system is selected these must be weighed in relation to the needs of the observer.

Reflectors are less expensive, size for size, than refractors, and they tend also to be more compact, especially with larger instruments. In a reflector, however, the mirror needs periodic recoating and realigning.

The design of a reflector means that air currents can be created inside the tube, resulting in slightly imperfect images. A refractor has a closed tube and so air currents cannot form, but light refracted by the lens can produce images displaying false colour fringes. This occurs because the lens cannot bring all the different wavelengths of light to a common focus.

The focal length of a telescope is the distance from the primary (lens or mirror) to the actual point of focus. Eyepieces, too, have different focal lengths, and magnification is altered simply by changing the eyepiece. To work out magnification, the focal length of the primary is divided by that of the eyepiece. In the case of a telescope of 60 cm (23.5 in), an eyepiece with a focal length of 2.5 cm (0.98 in) will give a magnification of 24 times.

The larger the focal length of the telescope, the higher the possible magnification will be. Refractors, with their long focal lengths, tend to be useful for lunar and planetary work, but the wide field of view obtained with reflectors makes them more suitable for the observation of nebulae and other objects in deep space.

All convex lenses produce an inverted image of a distant scene. In the photograph, *top*, the Moon can be seen as it would appear in a normal celestial telescope. For observation of the night sky, this inversion does not matter, as long as you remember that north is at the base of the image. In the photograph, *above*, the same image is seen as it would appear in a terrestrial telescope, or a celestial telescope fitted with an erecting prism. Such a prism introduces a series of additional air-to-glass surfaces into the light path and causes some loss of image brightness.

Refracting telescope

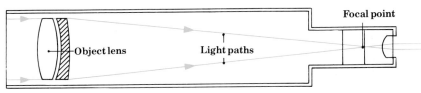

Focal point

Object lens   Light paths

Reflecting telescope   Light paths   Primary mirror

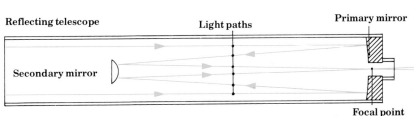

Secondary mirror

Focal point

Newtonian reflecting telescope

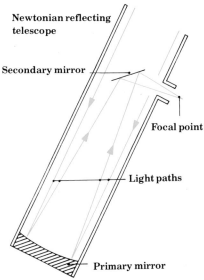

Secondary mirror

Focal point

Light paths

Primary mirror

Light enters
a refracting telescope through a lens and is bent to a focal point. The Cassegrain reflector uses a convex secondary mirror to reflect light back down and out through a hole in the primary

mirror. The Newtonian type of telescope uses a flat secondary, angled to reflect light out through the side of the tube. With some celestial objects, this can make the viewing angle awkward.

Caring for your reflector
If you keep your telescope indoors when not in use, it is important to let it cool down to the outside temperature before making any observations. This can take up to three hours if the telescope has been kept in a centrally heated room – which is not recommended. On taking the telescope outdoors, condensation may form on all exposed surfaces. A purpose-built observatory will prevent this problem (see pp 200–1).

When the telescope is not being used, always keep the mirrors covered to prevent dust and grease settling. If the mirror does require cleaning, remove the mirror cell and immerse it in water with completely dissolved soft washing powder. Remove the mirror after a few minutes and rinse it with distilled water. This is important, since ordinary tap water contains impurities that could pit the mirror's polished surface. Repeat this process if necessary and, after the final rinse, stand the mirror on its edge so that the water can run off naturally. Any stubborn droplets of water can be removed carefully with a piece of blotting paper.

Never try to remove dust or moisture from the mirror surface using a piece of cloth.

Telescope mounts and drives

Any instrument with a magnification of more than about 7 times must have some form of mounting if it is to be used to best effect. For small telescopes and more powerful types of binoculars (see pp 192–5), the altazimuth mount is commonly used. This permits movement of the telescope both vertically (altitude) and horizontally (azimuth).

One of the most popular and widely used altazimuth mounts is the Dobsonian. This is an ideal instrument for a beginner. It is freely portable and allows even relatively large telescopes to be handled easily.

The altazimuth mount is fine for low-power, wide-field work, but it becomes increasingly difficult to keep objects inside the narrow field of view of a large telescope, and both axes of the mount require constant adjustment.

To overcome this problem, a more advanced design is needed, and many astronomers use an equatorial mount, which permits an object to be followed by the adjustment of one axis only. The equatorial mount has two axes – polar and declination – placed at right angles to each other. The polar axis is permanently fixed, parallel to the Earth's axis and pointing to the celestial pole. Once the mount is set up, both axes are used to bring an object into the field of view. The declination axis is then clamped in position and the object followed by slowly turning the telescope tube around the polar axis.

Instead of turning mounts manually, they can be driven automatically. And although normally electrically driven, more sophisticated computer drives are now available.

Any type of mounting and drive system requires some routine maintenance. This includes lightly oiling any moving parts, lubricating bearings and checking components for wear and tear.

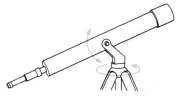

An altazimuth-mounted telescope moves in both horizontal and vertical planes. But because objects in the sky appear to move in a curved line, both axes have to be continually adjusted.

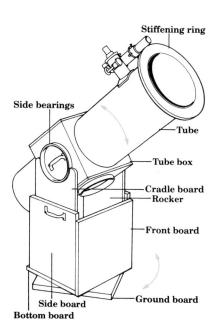

The Dobsonian mounting, *above*, is a do-it-yourself housing for a Newtonian telescope. The tube fits into a plywood box, which slots into a cradle. Two side bearings allow the box to pivot, and the cradle can turn freely on the baseboard.

Labels: Stiffening ring · Side bearings · Tube · Tube box · Cradle board · Rocker · Front board · Ground board · Side board · Bottom board

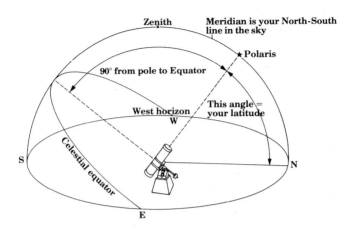

Zenith

Meridian is your North-South line in the sky

Polaris

90° from pole to Equator

West horizon
W

This angle = your latitude

S

Celestial equator

N

E

For an equatorial mount, adjust the polar axis to your own latitude, *above*. The mount must be perfectly level and the polar axis aimed at the celestial pole. With the telescope in place and secured, move the mount left or right, and the polar axis up and down, until the telescope points directly at the celestial pole.

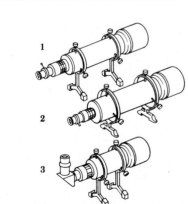

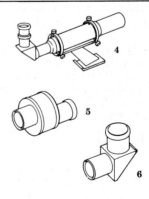

1

2

3

4

5

6

Viewfinders and prisms

In some amateur telescopes the angle of view is so narrow that locating an object in the night sky can be a problem. To overcome this, viewfinders are available, 1 and 2, which fit on the top of the telescope tube and present a wider field of vision. Right-angle finders, 3 and 4, can be helpful when viewing at awkward angles and erecting prisms, 5 and 6, are made for terrestrial observation.

The home observatory

Although small telescopes can easily be stored indoors and carried out to the observing site, larger instruments need a permanent observatory. Refractors above 10 cm (4 in) and reflectors above 20 cm (8 in) are strictly non-portable unless dismantled, which is not recommended.

The first requirement for a home observatory is a suitable site. A roof top should be avoided, since this will subject the telescope to an unwelcome mixture of vibration and warm air currents from the building beneath.

This reduces possible sites to the garden, where you should try to find an unobstructed horizon, taking into account trees, streetlights and so on. The form that the observatory should take can fall into one of three categories: the revolving dome, the run-off shed and the run-off roof. Although difficult to construct, the revolving dome offers maximum protection against stray light and the weather. For most reflectors, a run-off shed arrangement is suitable. It offers good protection against the weather when the telescope is not in use but is very exposed during observations.

Refractors and other tall telescopes are better housed in a run-off roof observatory. The high sides of such a building will not cause too much obstruction, since the height of the telescope will ensure that only a few degrees of sky will be lost from view.

Whatever type is chosen, the construction must be rigid. A good combination of materials is wood for the walls and clear, corrugated plastic for the roof. For long life and protection of your equipment, ensure that joints are waterproof.

Your observatory will require regular, routine maintenance. All moving parts must be kept lubricated and free from rust. Any exposed wooden surfaces should also be treated at least once a year with preservative or else be kept well painted.

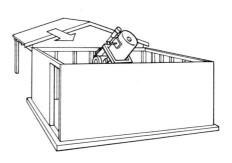

A sliding-roof observatory is reasonably simple to make from a garden shed. It offers good protection for the telescope but not for the observer, who is totally exposed in winter. The roof should be opened toward the north to give a clear view of southern skies, and to the south for northern skies.

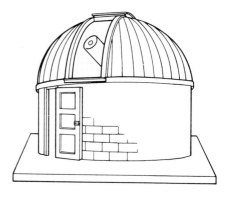

A domed-roof observatory allows a view of any part of the night sky. The walls of the structure can be made from treated wood and the dome, which must be light, reasonably strong and flexible, from sheets of aluminium.

All exterior wooden surfaces should be treated with preservative or kept well painted.

A hinged roof shutter is the simplest method of creating an observing space for the telescope.

A wall flap swings down when the telescope is in use.

The concrete base and rail are hidden by the structure.

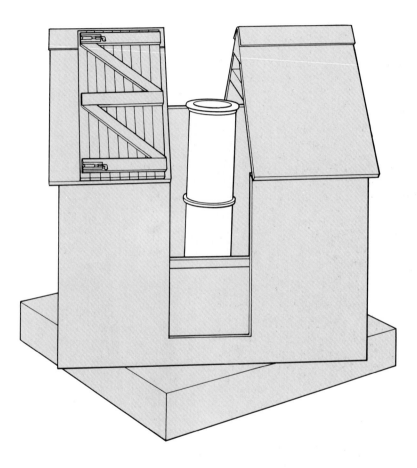

An observatory such as this should not cause too many constructional problems for a handy person. An ideal size is about 2.5 m (8 ft) square.

The base is made of concrete and the structure is built on a specially made iron rail mounted on wheels, which enables it to rotate.

Inside the observatory, keep clutter to a minimum. With some Newtonian reflectors, the eyepiece is on top of the tube, so you will need observing steps. If you have electric drive, a supply of electricity will be needed, but make sure that it is properly earthed. On a convenient inside wall, a mounted star chart will afford quick reference to many naked-eye stars.

Keeping records

Whether you are a casual star-gazer or a dedicated observer, it is a good idea to record any observations you make. There are different ways to do this. One is by means of a specially drawn-up sheet, which details all the relevant information of your sighting. Another is to keep a log book – especially if you are a regular observer. Whichever method is chosen, record as many details as possible while you are at the telescope; add any finishing touches while they are still fresh in your mind.

For those with a flair for drawing, a good sketch of the cloud belts of Jupiter, a lunar crater or the polar caps of Mars can equal or better a photographic record. Cameras can sometimes fail to capture subtle tones on planetary surfaces that the human eye can pick up quite clearly.

To make matters easier, it is advisable to prepare, on paper, blank discs for planetary observations. A good size has about a 5-cm (2-in) diameter, but those for Mercury could be smaller, while the opposite applies to Saturn— to accommodate the ring system. Solar observers should keep their disc drawings to a maximum of about 15 cm (6 in). When preparing discs for Jupiter, bear in mind that the planet has considerable polar flattening. Sketches of the lunar surface will have their dimensions governed by the size of the feature being drawn.

Drawing comets is something of an art in itself and practice is necessary before you can expect good results. Once seen, the comets' appearance as diffuse patches, rather than points of light, will give them away as they move against the fixed stars.

Whatever type of observation you make, try to record exactly what you see, rather than what you would like to see. Record, too, the date and time of any observations and also the clarity of the sky. The more these rules are adhered to, the more valuable your efforts will prove to be.

Antoniadi scale of seeing conditions

I Perfect seeing, without a quiver

II Slight undulations, with moments of calm lasting several seconds

III Moderate seeing, with large air tremors

IV Poor seeing, with constant troublesome undulations

V Bad seeing, scarcely allowing the making of a rough sketch

Seeing in this context is an estimate of the steadiness of the image. The scale, *above,* is universal, and it will help when comparing observations of the same object made from different sites and under different conditions.

Record cards are vital for the serious observer. Their standard format and layout ensure that the same type of information is always recorded. This is particularly important when tracking some of the more transient planetary features.

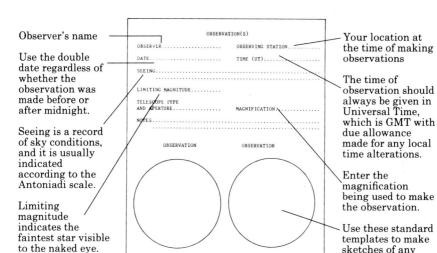

Observer's name

Use the double date regardless of whether the observation was made before or after midnight.

Seeing is a record of sky conditions, and it is usually indicated according to the Antoniadi scale.

Limiting magnitude indicates the faintest star visible to the naked eye.

Any additional notes

Your location at the time of making observations

The time of observation should always be given in Universal Time, which is GMT with due allowance made for any local time alterations.

Enter the magnification being used to make the observation.

Use these standard templates to make sketches of any observed features.

OBSERVATION(S)

OBSERVER...................... OBSERVING STATION..........
DATE....................... TIME (UT)................
SEEING...
LIMITING MAGNITUDE..........
TELESCOPE TYPE
AND APERTURE................. MAGNIFICATION..............
NOTES...
....

OBSERVATION   OBSERVATION

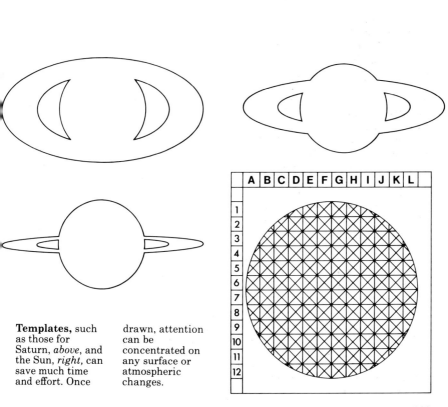

	A	B	C	D	E	F	G	H	I	J	K	L	
1													
2													
3													
4													
5													
6													
7													
8													
9													
10													
11													
12													

Templates, such as those for Saturn, *above,* and the Sun, *right,* can save much time and effort. Once drawn, attention can be concentrated on any surface or atmospheric changes.

203

Appendices

Future positions of the planets from 1985 to 2000

The positions of Mercury and Venus are given by quoting their elongation, east or west of the Sun; east means the planet is visible before dawn, west in the evening. The month of first appearance at a given elongation is stated.

The bright outer planets – Mars, Jupiter and Saturn – are given by reference to constellation only. Where no month or constellation is named, the planet is too close to the Sun to be visible.

	Mercury Magnitude −1.8	Venus Magnitude −4.4	Mars Magnitude −2.8
1985	W Jan, April, Aug, Dec E March, June, Oct	W April E Jan	Jan: Pisces April–June: Taurus Sept: Leo; Oct–Dec: Virgo
1986	W Jan, April, July E Feb, May, Sept, Nov	W Nov E Jan	Jan–March: Libra April–Oct: Sagittarius Oct–Nov: Capricornus Dec: Aquarius .
1987	W March, July, Nov E Jan, May, Sept	W Jan E Sept	Jan: Aquarius Feb–March: Pisces March: Aries April–May: Taurus June: Gemini Oct–Dec: Virgo Dec: Leo
1988	W Feb, June, Oct E Jan, April, Aug, Dec	W June E Jan	Jan: Libra Jan–Feb: Ophiuchus March: Sagittarius April: Capricornus May–July: Aquarius
1989	W Feb, May, Oct E Jan, April, July, Nov	W Jan E April	Jan: Pisces; Jan–Feb: Aries Feb–April: Taurus May–June: Gemini June–July: Cancer July: Leo; Nov: Virgo Dec: Libra
1990	W Jan, May, Sept E March, July, Nov	W Jan E Nov	Jan: Ophiuchus Feb–March: Sagittarius March: Capricornus April–May: Aquarius May–July: Pisces July–Aug: Aries Aug–Dec: Taurus
1991	W Jan, April, Aug, Dec E March, June, Oct	W Jan E Sept	Jan–March: Taurus April–May: Gemini May–July: Cancer July–Aug: Leo; Sept: Virgo
1992	W Jan, March, Aug, Nov E Feb, June, Sept	W Jan E June	Jan–Feb: Sagittarius Feb–March: Capricornus March–May: Aquarius May–June: Pisces June–July: Aries July–Sept: Taurus Sept–Nov: Gemini Nov–Dec: Cancer

Uranus, Neptune and Pluto move very slowly in the sky, and the months given state when the planet should be visible. The positions given are for mid-year; each planet is moving eastward and will be close to the positions given. It is not possible to observe the planets satisfactorily in a telescope when they are close to the observer's horizon. In some of the positions quoted, the planets are visible only to observers in the northern hemisphere; in others, only observers in the southern hemisphere can see them.

Jupiter Magnitude −2.5	Saturn Magnitude −0.4	Uranus Magnitude 6	Neptune Magnitude 7.7	Pluto Magnitude 13.7
Feb: Sagittarius Feb–Dec: Capricornus	Jan–Oct, Dec: Libra	Jan–Nov: NW of θ Ophiuchi	Jan–Nov: W of μ Sagittarii	Jan–Oct, Dec: ENE of τ Virginis
March–Dec: Aquarius	Jan–Feb: Scorpius Feb–May: Ophiuchus June–Nov: Scorpius Dec: Ophiuchus	Jan–Nov: N of θ Ophiuchi	Jan–Nov: N of λ Sagittarii	Jan–Oct, Dec: E of τ Virginis
Jan: Aquarius, Pisces March–Dec: Pisces	Jan–Oct, Dec: Ophiuchus	Jan–Nov: NE of θ Ophiuchi	Jan–Nov: NNE of λ Sagittarii	Jan–Oct, Dec: Close to 109 Virginis
Jan–March, April–June: Aries July–Dec: Taurus	Jan–Nov: Sagittarius	Jan–Nov: SW of μ Sagittarii	Jan–Nov: NE of λ Sagittarii	Jan–Oct, Dec: SE of 109 Virginis
Jan–April, June–July, Taurus Aug–Dec: Gemini	Jan (late)–Dec: Sagittarius	Jan–Nov: S of μ Sagittarii	Jan–Nov: SW of ξ Sagittarii	Jan–Oct, Dec: S of 110 Virginis
Jan–June: Gemini Aug–Dec: Cancer	Feb–Dec: Sagittarius	Jan–Nov: NE of λ Sagittarii	Jan–Nov: Close to o Sagittarii	Jan–Oct, Dec: SE of 110 Virginis
Jan–July: Cancer Sept–Dec: Leo	Feb–Dec: Capricornus	Jan–Nov: N of φ Sagittarii	Jan–Nov: Just N of o Sagittarii	Jan–Oct, Dec: NNW of μ Serpentis Caput
Jan–Aug: Leo Oct–Dec: Virgo	Feb–Dec: Capricornus	Jan–Nov: Close to o Sagittarii	Jan–Nov: SE of π Sagittarii	Jan–Oct, Dec: SW of μ Serpentis Caput

	Mercury Magnitude −1.8	Venus Magnitude −4.4	Mars Magnitude −2.8
1993	W Jan, March, July Nov E Feb, May, Sept	W April E Jan	Jan–April: Gemini April–May: Cancer May–Aug: Leo Aug–Sept: Virgo Oct–Nov: Libra
1994	W Feb, June, Oct E Jan, May, Aug, Dec	W Nov E Jan	March: Aquarius April–May: Pisces May–June: Aries July–Aug: Taurus Aug–Sept: Gemini Oct–Nov: Cancer Nov–Dec: Leo
1995	W Feb, June, Oct E Jan, April, Aug, Nov	W Jan E Aug	Jan–March: Leo March: Cancer April–July: Leo July–Sept: Virgo Sept–Oct: Libra Nov: Ophiuchus Dec: Sagittarius
1996	W Jan, May, Sept E Jan, April, July, Nov	W June E Jan	Jan: Capricornus May: Aries June–July: Taurus July–Sept: Gemini Sept–Oct: Cancer Oct–Dec: Leo
1997	W Jan, May, Sept, Dec E Jan, March, July Oct	W Jan E April	Jan–Aug: Virgo Aug–Sept: Libra Sept–Nov: Ophiuchus Nov–Dec: Sagittarius Dec: Capricornus
1998	W Jan, April, Aug, Dec E Feb, June, Oct	W Jan E Jan, Nov	Jan: Capricornus Jan–Feb: Aquarius Feb: Pisces Aug: Gemini Aug–Sept: Cancer Sept–Nov: Leo Nov–Dec: Virgo
1999	W Jan, March, July, Nov E Feb, June, Sept	W Aug E Jan	Jan–Feb: Virgo Feb–April: Libra April–July: Virgo July–Aug: Libra Aug–Sept: Scorpius Sept–Oct: Ophiuchus Oct–Nov: Sagittarius Nov–Dec: Capricornus
2000	W March, July, Nov E Jan, May, Aug	W Jan E June	Jan: Capricornus, Aquarius Feb–March: Pisces March–April: Aries May: Taurus Aug: Cancer Sept–Oct: Leo Nov: Virgo

Jupiter Magnitude -2.5	Saturn Magnitude -0.4	Uranus Magnitude 6	Neptune Magnitude 7.7	Pluto Magnitude 13.7
Jan–Sept, Nov–Dec: Virgo	Jan, March Capricornus April–Dec: Aquarius	Jan–Nov: SE of π Sagittarii	Jan–Nov: E of π Sagittarii	Jan–Oct, Dec: SSW of μ Serpentis Caput
Jan–Oct, Dec: Libra	Jan, March–Dec: Aquarius	Jan–Nov: N of ω Sagittarii	Jan–Nov: N of ω Sagittarii	Jan–Oct, Dec: SE of μ Serpentis Caput
Jan: Scorpius Feb–Nov: Ophiuchus	Jan–Feb, April–Dec: Aquarius	Jan–Nov: SW of β Capricornii	Jan–Nov: NNE of ω Sagittarii	Jan–Oct, Dec: WSW of δ Ophiuchi
Jan–Dec: Sagittarius	Jan: Aquarius March–Dec: Pisces	Jan–Nov: SE of β Capricornii	Jan–Nov: N of 62 Sagittarii	Jan–Oct, Dec: SW of δ Ophiuchi
Jan (end)–Dec: Capricornus	Jan–Feb, April–Dec: Pisces	Jan–Nov: Close to ν Capricorni	Jan–Nov: SW of β Capricorni	Jan–Oct, Dec: SE of ε Ophiuchi
Jan: Capricornus March–Dec: Aquarius	Jan–March: Pisces May–Dec: Aries	Jan–Nov: NW of θ Capricorni	Jan–Nov: SSW of β Capricorni	Jan–Oct, Dec: Close to ν Ophiuchi
Jan–Feb, April–Dec: Pisces	Jan–mid April, June–Dec: Aries	Jan–Nov: NE of θ Capricorni	Jan–Nov: SSE of β Capricorni	Jan–Oct, Dec: Close to ζ Ophiuchi
Jan: Pisces Feb–April: Aries June–Dec: Taurus	Jan–April, June–Nov: Aries Dec: Taurus	Jan–Nov: NW of γ Capricorni	Jan–Nov: Close to ρ Capricorni	Jan–Oct, Dec: Close to 20 Ophiuchi

Units in Astronomy	Kilometres	Miles	Astronomical units
Astronomical unit (a.u.)– distance Earth to Sun	1.496×10^8	9.296×10^7	
Light-year (l.y.)	9.461×10^{12}	5.879×10^{12}	63×10^3
Parsec (pc)	30.857×10^{12}	19.174×10^{12}	206×10^3

Kiloparsec (kpc) = 10^3 parsecs
Megaparsec (mpc) = 10^6 parsecs
Velocity of light in a vacuum = 2.998×10^5 km per second (km/s)

Powers of 10

$10^0 = 1$
$10^2 = 100$
$10^3 = 1,000$
$10^4 = 10,000$
$10^5 = 100,000$
$10^6 = 1,000,000$ (one million)
$10^9 = 1,000,000,000$ (one billion)
$10^{12} = 1,000,000,000,000$ (one trillion)

$10^{-1} = 0.1$
$10^{-2} = 0.01$
$10^{-3} = 0.001$ (one-thousandth)
$10^{-6} = 0.000\,001$ (one-millionth)

Ratio of brightness of celestial objects

Difference in magnitude		Multiple by which brightness differs	
0.1	0.9	1.10	2.29
0.2	1.0	1.20	2.51
0.25	1.5	1.26	3.98
0.3	2.0	1.32	6.31
0.4	2.5	1.45	10.00
0.5	3.0	1.58	15.85
0.6	4.0	1.74	39.81
0.7	5.0	1.91	100.00
0.75	10.0	2.00	10,000.00
0.8	15.0	2.09	1,000,000.00

Astronomical symbols

The Zodiac

♈	Aries
♉	Taurus
♊	Gemini
♋	Cancer
♌	Leo
♍	Virgo
♎	Libra
♏	Scorpio
♐	Sagittarius
♑	Capricornus
♒	Aquarius
♓	Pisces

The Solar System

☉	Sun
☿	Mercury
♀	Venus
⊕	Earth
♂	Mars
♃	Jupiter
♄	Saturn
♅	Uranus
♆	Neptune
♇	Pluto
☽	Moon

Measurement of angles

°	degrees
′	minutes
″	seconds

Astrophysics

m (or mag)	Apparent magnitude
M	Absolute magnitude
m_v	Apparent visual magnitude
m_{pg}	Apparent photographic magnitude
λ	Wavelength
Å	Ångstrom unit = 10^{-7} mm
μ	Micron = 10^{-3} mm

Conversions

Mass

1 gram (g)	=	0.035273 ounces
1 kilogram (kg)	=	2.20462 pounds
1 ounce (oz)	=	28.3495 grams
1 pound (lb)	=	0.453592 kilograms

Length

1 millimetre (mm)	=	0.03937 inches
1 centimetre (cm)	=	0.393701 inches
1 metre (m)	=	3.28084 feet
1 kilometre (km)	=	0.621371 miles
1 inch (in)	=	25.400 millimetres
1 foot (ft)	=	0.304799 metres
1 yard (yd)	=	0.914399 metres
1 mile (ml)	=	1.609344 kilometres

Supplementary data

Light-time for travelling one astronomical unit	8.3 minutes
Average distance Earth to Moon	384,400 km
Length of the year:	
Tropical (equinox to equinox)	365.24219 days
Sidereal (fixed star to fixed star)	365.25636 days
Anomalistic (apse to apse)	365.25964 days
Eclipse (Moon's node to Moon's node)	346.62003 days
Length of the month:	
Tropical (equinox to equinox)	27.32158 days
Sidereal (fixed star to fixed star)	27.32166 days
Anomalistic (apse to apse)	27.55455 days
Draconic (node to node)	27.21222 days
Synodic (New Moon to New Moon)	29.53059 days

Length of the day:

Mean solar day $24^h 03^m 56^s.555 = 1^d.00273791$ mean sidereal time

Mean sidereal day $23^h 56^m 04^s.091 = 0^d.99726957$ mean solar time

Sidereal rotation period of the Earth

         $23^h 56^s 04.099 = 0^d.99726966$ mean solar time

Figure of the Earth:	
Equatorial radius	6,378.140 km
Polar radius	6,356.755 km
Mass of the Earth	5.9742×10^{24} kg
Mass of the Moon	7.3483×10^{22} kg
Mass of the Sun	1.989×10^{30} kg
The Galactic System:	
Pole of galactic plane	$\alpha 12^h 49^m.0$, $\delta + 27° 24'$
Point of zero longii	$\alpha 17^h 42^m.4$, $\delta - 28° 55'$
Galactic longitude of North Celestial Pole	$123°.00/s$
Solar motion with respect to bright stars	20.0 km/s
Period of revolution of Sun about centre of the Galaxy	2.2×10^8 yr
Conversion factors:	
Light-year	9.4607×10^{12} km $= 63,240$ a.u. $= 0.30660$ pc
Parsec	30.857×10^{12} km $= 206,265$ a.u. $= 3.2616$ l.y.

Limiting magnitudes of binoculars and telescopes

Aperture of instrument	7×50 mm Binoculars	7.5 cm (3 in) Telescope	10 cm (4 in) Telescope	15 cm (6 in) Telescope	20 cm (8 in) Telescope	30 cm (12 in) Telescope
Closest separable stars (sec)	$4''.56$	$1''.28$	$1''.52$	$0''.76$	$0''.57$	$0''.38$
Approximate Magnitude limit	9.0	11.4	12.0	12.9	13.5	14.4

Temperature

To convert:

Fahrenheit (°F) to Centigrade (°C) subtract 32 and multiply by $\frac{5}{9}$

Centigrade (°C) to Fahrenheit (°F) multiply by $\frac{9}{5}$ and add 32

Centigrade (°C) to Kelvin (K) add 273.155

Solar eclipses for the years 1985–2000

Date	Eclipse visible	Type
1985 May 19	Arctic	Partial
1985 November 12	South Pacific, Antarctica	Total
1986 April 9	Antarctic	Partial
1986 October 3	North Atlantic	Total
1987 March 29	Argentina, Atlantic, Central Africa, Indian Ocean	Total
1987 September 23	USSR, China, Pacific	Annular
1988 March 18	Indian Ocean, East Indies, Pacific	Total
1989 March 7	Arctic	Partial
1989 August 31	Antarctic	Partial
1990 January 26	Antarctic	Annular
1990 July 22	Finland, USSR, Pacific	Total
1991 January 15–16	Australia, New Zealand, Pacific	Annular
1991 July 11	Pacific, Mexico, Brazil	Total
1992 January 4–5	Central Pacific	Annular
1992 June 30	South Atlantic	Total
1992 December 24	Arctic	Partial
1993 May 21	Arctic	Partial
1993 November 13	Antarctic	Partial
1994 May 10	Pacific, Mexico, USA, Canada	Annular
1994 November 3	Peru, Brazil, South Atlantic	Total
1995 April 29	South Pacific, Peru, South Atlantic	Annular
1995 October 24	Iran, India, East Indies, Pacific	Total
1996 April 17	Antarctic	Partial
1996 October 12	Arctic	Partial
1997 March 9	USSR, Arctic	Total
1997 September 2	Antarctic	Partial
1998 February 26	Pacific, Atlantic	Total
1998 August 22	Indian Ocean, East Indies, Pacific	Annular
1999 February 16	Indian Ocean, Australia, Pacific	Annular
1999 August 11	Atlantic, British Isles, France, Turkey, India	Total
2000 February 5	Antarctic	Partial
2000 July 1	Pacific Ocean, South America	Partial
2000 July 31	Asia, Alaska, Canada, Arctic, Greenland	Partial
2000 December 25	North America, Gulf of Mexico, South Caribbean Sea, Greenland	Partial

Lunar eclipses for the years 1985–2000

Date	Magnitude (% in shadow)	Date	Magnitude (% in shadow)
1985 May 14	Total (100)	1992 December 10	Total
1985 October 28	Total	1993 June 4	Total
1986 April 24	Total	1993 November 29	Total
1986 October 17	Total	1994 May 25	28
1987 October 7	1	1995 April 15	12
1988 August 27	30	1996 April 4	Total
1989 February 20	Total	1996 September 27	Total
1989 August 17	Total	1997 March 24	93
1990 February 9	Total	1997 September 16	Total
1990 August 6	68	1999 July 28	42
1991 December 21	9	2000 January 21	Total
1992 June 15	69	2000 July 16	Total

Satellites of the planets

Planet	Name of satellite	Average distance from planet (000 km)	Sidereal period (days)	Inclination to planet's orbit (degrees)	Orbital eccentricity	Diameter (km)	Mass (kg)
Mars	Phobos	9.38	0.31891	1.0	0.018	14×10*	9.6×10^{15}
	Deimos	23.5	1.26244	2.0	0.002	8×6*	2.0×10^{15}
Jupiter	Metis	128.2	0.294	0**	0**	40**	Not known
	Adrastea	128.5	0.297	0**	0**	30**	Not known
	Amatthea	181.3	0.489	0.455	0.003	240	Not known
	Thebe	223	0.675	0**	0**	80**	Not known
	Io	412.6	1.769	0.027	0.000	3,632	8.916×10^{22}
	Europa	670.9	3.551	0.468	0.000	3,126	4.873×10^{22}
	Ganymede	1,070	7.155	0.183	0.001	5,276	1.490×10^{23}
	Callisto	1,880	16.689	0.253	0.007	4,820	1.490×10^{23}
	Leda	11,110	240	27	0.147	2–14	Not known
	Himalia	11,470	250.6	28	0.158	170	Not known
	Lysithea	11,710	260	29	0.12	6–32	Not known
	Elara	11,740	260.1	26	0.207	80	Not known
	Ananke	20,700	617	147	0.169	6–28	Not known
	Carme	22,350	692	163	0.207	8–40	Not known
	Pasiphae	23,300	735	147	0.40	8–46	Not known
	Sinope	23,700	758	156	0.275	6–36	Not known
Saturn	1980 S28	137.67	0.602	0**	0**	20×10*	Not known
	1980 S27	139.35	0.613	0**	0**	70×40*	Not known
	1980 S26	141.7	0.629	0**	0**	55×35*	Not known
	1980 S3	151.422	0.694	0**	0**	70×50*	Not known
	1980 S1	151.472	0.695	0**	0**	110×80*	Not known
	Mimas	185.54	0.942	1.517	0.020	392	4.5×10^{19}**
	Enceladus	238.04	1.370	0.023	0.004	500	8.4×10^{19}
	Tethys	294.67	1.888	1.093	0.000	1,000	7.55×10^{20}
	1980 S13	294.67	1.888	1.0**	0**	17×13*	Not known
	1980 S25	294.67	1.888	1.0**	0**	17×11*	Not known
	Dione	377.42	2.737	0.023	0.002	1,120	1.05×10^{21}
	1980 S6	378.06	2.739	0**	0**	18×15*	Not known
	Rhea	527.1	4.518	0.35	0.001	1,530	2.49×10^{21}
	Titan	1,221.86	15.945	0.33	0.029	5,150	1.35×10^{23}
	Hyperion	1,481	21.277	0.4	0.104	205×110*	Not known
	Iapetus	3,560.8	79.331	14.7	0.028	1,460	1.88×10^{21}
	Phoebe	12,954	550.4	150	0.163	220	Not known
Uranus	Miranda	130	1,414	3.4	0.000	320**	2.2×10^{19}**
	Ariel	192	2.520	0	0.003	1,330	1.6×10^{21}**
	Umbriel	267	4.144	0	0.004	1,110	9.3×10^{20}**
	Titania	438	8.706	0	0.002	1,600	2.8×10^{21}**
	Oberon	587	13.463	0	0.001	1,630	2.9×10^{21}**
Neptune	Triton	355	5.877	160	0.00	3,200**	3.4×10^{22}
	Nereid	5,562	365.21	28	0.75	300**	2.8×10^{19}**
Pluto	Charon	19.7	6.387	94	0.0	1,500**	1.6×10^{21}**

*These are not diameters; the satellites are elongated and the values give their two dimensions.
**Values uncertain.

The Constellations

English name	Latin name	Genitive	Abbreviation
Andromeda	Andromeda	Andromedae	And
Pump	Antlia	Antliae	Ant
Bird of Paradise	Apus	Apodis	Aps
Water Bearer	Aquarius	Aquarii	Aqr
Eagle	Aquila	Aquilae	Aql
Altar	Ara	Area	Ara
Ram	Aries	Arietis	Ari
Charioteer	Auriga	Aurigae	Aur
Herdsman	Boötes	Boötis	Boo
Chisel	Caelum	Caeli	Cae
Giraffe	Camelopardalis	Camelopardalis	Cam
Crab	Cancer	Cancri	Cnc
Hunting Dogs	Canes Venatici	Canum Venaticorum	CVn
Big Dog	Canis Major	Canis Majoris	CMa
Little Dog	Canis Minor	Canis Minoris	CMi
Sea Goat	Capricornus	Capricorni	Cap
Ship's Keel**	Carina	Carinae	Car
Cassiopeia	Cassiopeia	Cassiopeiae	Cas
Centaur	Centaurus	Centauri	Cen
Cepheus	Cepheus	Cephei	Cep
Whale	Cetus	Ceti	Cet
Chameleon	Chamaeleon	Chamaeleonis	Cha
Compass	Circinus	Circini	Cir
Dove	Columba	Columbae	Col
Berenice's Hair	Coma Berenices	Comae Berenices	Com
Southern Crown	Corona Australis	Coronae Australis	CrA
Northern Crown	Corona Borealis	Coronae Borealis	CrB
Raven	Corvus	Corvi	Crv
Cup	Crater	Crateris	Crt
Southern Cross	Crux	Crucis	Cru
Swan	Cygnus	Cygni	Cyg
Dolphin	Delphinus	Delphini	Del
Swordfish	Dorado	Doradus	Dor
Dragon	Draco	Draconis	Dra
Little Horse	Equuleus	Equulei	Equ
River Eridanus	Eridanus	Eridani	Eri
Furnace	Fornax	Fornacis	For
Twins	Gemini	Geminorum	Gem
Crane	Grus	Gruis	Gru
Hercules	Hercules	Herculis	Her
Clock	Horologium	Horologii	Hor
Hydra (Water Monster)	Hydra	Hydrae	Hya
Sea Serpent	Hydrus	Hydri	Hyi
Indian	Indus	Indi	Ind
Lizard	Lacerta	Lacertae	Lac
Lion	Leo	Leonis	Leo
Little Lion	Leo Minor	Leonis Minoris	LMi
Hare	Lepus	Leporis	Lep
Scales	Libra	Librae	Lib
Wolf	Lupus	Lupi	Lup
Lynx	Lynx	Lyncis	Lyn
Harp	Lyra	Lyrae	Lyr

English name	Latin name	Genitive	Abbreviation
Table (Mountain)	Mensa	Mensae	Men
Microscope	Microscopium	Microscopii	Mic
Unicorn	Monoceros	Monocerotis	Mon
Fly	Musca	Muscae	Mus
Level (Square)	Norma	Normae	Nor
Octant	Octans	Octantis	Oct
Ophiuchus (Serpent Bearer)	Ophiuchus	Ophiuchi	Oph
Orion	Orion	Orionis	Ori
Peacock	Pavo	Pavonis	Pav
Pegasus (Winged Horse)	Pegasus	Pegasi	Peg
Perseus	Perseus	Persei	Per
Phoenix	Phoenix	Phoenicis	Phe
Easel	Pictor	Pictoris	Pic
Fish	Pisces	Piscium	Psc
Southern Fish	Piscis Austrinus	Piscis Austrini	PsA
Ship's Stern**	Puppis	Puppis	Pup
Ship's Compass**	Pyxis	Pyxidis	Pyx
Net	Reticulum	Reticuli	Ret
Arrow	Sagitta	Sagittae	Sge
Archer	Sagittarius	Sagittarii	Sgr
Scorpion	Scorpius	Scorpii	Sco
Sculptor	Sculptor	Sculptoris	Scl
Shield	Scutum	Scuti	Sct
Serpent	Serpens	Serpentis	Ser
Sextant	Sextans	Sextantis	Sex
Bull	Taurus	Tauri	Tau
Telescope	Telescopium	Telescopii	Tel
Triangle	Triangulum	Trianguli	Tri
Southern Triangle	Triangulum Australe	Trianguli Australis	TrA
Toucan	Tucana	Tucanae	Tuc
Big Bear	Ursa Major	Ursae Majoris	UMa
Little Bear	Ursa Minor	Ursae Minoris	UMi
Ship's Sails**	Vela	Velorum	Vel
Virgin	Virgo	Virginis	Vir
Flying Fish	Volans	Volantis	Vol
Little Fox	Vulpecula	Vulpeculae	Vul

**Formerly grouped to form the constellation Argo Navis, the Argonauts' ship.

Important annual meteor showers

Name	Total duration	Maximum number of meteors	Average per hour
Quadrantids	January 1–6	January 3	80
Lyrids	April 19–24	April 22	15
η Aquarids	May 2–7	May 4	40
α Scorpiids	April 28–May 12	April 28–May 12	20
δ Aquarids	July 15–August 15	July 28	20
Perseids	July 27–August 17	August 12	75
Orionids	October 12–16	October 21	20
Taurids	October 26–November 25	November 4	12
Leonids	November 15–19	November 17	10
Geminids	December 7–15	December 14	60

Some star clusters

Name of constellation	Object	RA (hrs min)		Dec. (° ')		Type of cluster	Magni-tude	Remarks
Auriga	M38	05	25	+35	48	Open	7.4	Visible with binoculars
Auriga	M37	05	49	+32	36	Open	6.2	Visible with binoculars
Cancer (Praesepe or The Beehive)	M44	08	37	+20	12	Open	3.7	Visible to naked eye
Canes Venatici	M3	13	40	+28	36	Globular	6.4	Visible with binoculars
Cassiopeia	M103	01	30	+60	24	Open	7.4	Visible with binoculars
Centaurus	NGC 3766	11	34	−61	18	Open	5.1	Visible with binoculars
Centaurus ($\frac{1}{2}°$ diameter)	ω	13	24	−47	0	Globular	3.7	Visible to naked eye
Crux	NGC 4755	12	51	−60	6	Open	5.2	Visible with binoculars
Cygnus	M39	21	30	+48	12	Open	5.2	Visible with binoculars
Gemini	M35	06	06	+24	24	Open	5.3	Visible with binoculars
Hercules (Great Cluster)	M13	16	40	+36	36	Globular	5.7	Visible with binoculars
Pegasus	M15	21	28	+12	0	Globular	6.0	Visible with binoculars
Perseus (Swordhandle or Double Cluster)	NGC 869	02	18	+56	54	Open	4.4	Visible to naked eye
Perseus	NGC 884	02	18	+56	54	Open	4.7	Visible to naked eye
Perseus	M34	02	39	+42	30	Open	5.5	Visible with binoculars
Sagittarius	M23	17	54	−19	0	Open	6.9	Visible with binoculars
Scorpius	M6	17	37	−32	12	Open	5.3	Visible with binoculars
Scorpius	M7	17	51	−34	48	Open	3.2	Visible with binoculars
Scutum	M11	18	48	−06	48	Open	6.3	Visible to naked eye
Taurus (The Pleiades)	M45	03	44	+24	0	Open	1.6	Visible to naked eye
Triangulum Australe	NGC 6025	15	59	−60	24	Open	5.8	Visible with binoculars
Tucana (Star 47 Tucanae)	NGC 104	00	22	−72	24	Globular	3.0	Visible to naked eye

Some variable stars visible without a telescope

Name	RA (hrs min)		Dec. (° ')		Max. mag.	Min. mag.	Period (days)	Type
γ Cassiopeiae	00	54	+60	27	1.6	3.0	—	Irregular
σ Ceti	02	17	−03	12	2.0*	10.1	331.48	Long period star
β Persei	03	05	+40	46	2.2	3.5	2.8673	Eclipsing binary of Algol type
α Orionis	05	53	+07	24	0.4	1.3	2070	Semi-regular
α Herculis	17	12	+14	27	3.0	4.0	100	Semi-regular
β Lyrae	18	48	+33	18	3.4	4.3	12.9080	Type star of Lyrae class of variable
μ Cephei	21	42	+58	33	3.6	5.1	—	Semi-regular
δ Cephei	22	27	+58	10	3.9	5.0	5.3663	Type star of Cepheid class of variable
ρ Cassiopeiae	23	52	+57	13	4.1	6.2	—	Irregular

*Telescope required at maximum magnitude.

Societies and organizations

Britain
British Astronomical Association,
Burlington House, Piccadilly,
London, W1V 0NL
Royal Astronomical Society,
Burlington House, Piccadilly,
London W1V 0NL
Royal Meteorological Society,
James Glaisher House,
Grenville Place, Bracknell,
Berks RG12 1BX
America
**American Association of Variable
Star Observers** (AAVSO),
187 Concord Avenue, Cambridge,
Massachusetts 02138
American Meteorological Society,
45 Beacon Street, Boston,
MA 03108
**National Oceanic and Atmospheric
Administration** (NOAA), National
Weather Service, 8060 13th Street,
Silver Spring, MD 20910
Australia
British Astronomical Association
(New South Wales Branch),

Sydney Observatory, Sydney,
New South Wales, 2001
Royal Meteorological Society,
Australian Branch,
Meteorological Department,
University of Melbourne,
Parkville, Victoria
Canada
**Canadian Meteorological and
Oceanographic Society,** Suite 805,
151 Slater Street, Ottawa,
Ontario, K1P 5H3
**Royal Astronomical Society of
Canada,** 136 Dupont Street, Toronto,
Ontario, M5R 1VS
South Africa
**Astronomical Society of Southern
Africa,** c/o South African
Astronomical Observatory,
P.O. Box 9, Observatory, 7935, Cape
New Zealand
**Royal Astronomical Society of New
Zealand,** P.O. Box 3181,
Wellington C1
New Zealand Meteorological Society
P.O. Box 3263, Wellington

Glossary

Terms which are referred to in the text without a full explanation are included in the glossary. Some other terms which are clearly defined in the book are not included.

A

Altazimuth mount (Dobsonian mount): a type of mount enabling a telescope to move independently both horizontally (in azimuth) and vertically (altitude). It necessitates frequent adjustment (see *Equatorial mount*).

Anticyclone: a centre of high pressure from which winds spiral outward over a wide area. In temperate zones it usually moves slowly toward the east.

Antisolar point: the notional point lying on an extension of the line that joins the Sun to an observer's eye.

Aphelion: the farthest point of an orbiting body from the Sun (see *Perihelion*).

Arc second: the unit used to measure angular distance across the sky; it is $\frac{1}{3600}$ of a degree (one degree $=\frac{1}{360}$ of a circle).

Asteroid: a small planetary body, or 'planetesimal', orbiting the Sun, usually between the orbits of Mars and Jupiter.

Asthenosphere: a hot, pliable layer lying below the crust and mantle of the Moon.

Astronomical unit (a.u.): a unit of length defined by the average radius of the Earth's orbit; it measures 149,597,870 km (92,960,116 mls).

Atmosphere: the gaseous shell surrounding a planet or major moon; the outer layers of a star. Weather events take place in the lower layers of the Earth's atmosphere.

Auroral ovals: zones lying around the magnetic poles above which aurorae are most often seen, due to the presence of charged particles in the upper atmosphere.

Axial inclination: the angle between the polar axis of a planet and the plane of its orbit around its parent star.

Axial rotation: the amount of time it takes for a planet to rotate once about its axis.

C

Celestial equator: the projection of the Earth's Equator on the celestial sphere.

Celestial pole: the projection of the Earth's polar axis on the celestial sphere.

Celestial sphere: an imaginary sphere, centred on the Earth, used for plotting the coordinates of celestial bodies.

Chromosphere: the part of the Sun's atmosphere that lies below the corona and above the Sun's surface, or *photosphere*.

Circumzenithal arc: a short, coloured halo arc centred on the zenith and appearing close to the top of the 46° halo (see *Halo*).

Clusters: one of two categories of groups of stars. 1 Open clusters do not have a definite shape and usually contain dust and gas as well as several hundred stars. They are located in the spiral arms of the Galaxy. 2 Globular clusters are concentrated groups of thousands of stars, forming a shape. They are more distant stars, positioned in a spherical envelope near the edge of our Galaxy.

Convection: the process whereby hot air rises and cold air sinks, creating a circulation.

Cyclone: a centre of low pressure into which winds spiral counter-clockwise in the northern hemisphere and clockwise in the southern. A tropical cyclone is a revolving storm, the most intense being known also as hurricanes or typhoons.

D

Declination (dec): the equivalent on the celestial sphere of terrestrial latitude. It is measured in degrees North (positive) or South (negative) from the celestial equator (0°) to the poles (90°). The angle between the celestial equator and a celestial body gives the declination of that body (see *Right Ascension*).

Depression: a region of low pressure. The alternative term for an extratropical, or temperate, cyclone.

Double star system: two stars in orbit around each other that revolve around a common centre of gravity. An optical double star system consists of a pair of stars that appear close only because they lie in almost the same line of sight.

E

Eclipse: this results in the image of either the Sun or the Moon as seen from Earth being partially or totally obscured. A solar eclipse is a form of occultation, when the Moon passes in front of the Sun. During a lunar eclipse, the Earth's shadow falls across the Moon.

Eclipsing binary: a type of binary star in which the orbit of both components appears edge-on from the Earth. As a result, these components alternately eclipse each other so that the apparent magnitude of the binary pair is constantly changing.

Ellipse: a regular, oval-shaped loop, constructed by cutting obliquely through a cone; hence elliptical orbit, elliptical galaxy, etc. Planets follow elliptical orbits, with the Sun at one focus of the ellipse.

Equatorial mount: a type of mount that permits a telescope to rotate on an axis parallel to the Earth's axis and also at right angles to it. The motion of a body can be followed by the adjustment of only one axis (see *Altazimuth mount*).

Equinox: each of two days in the year when both day and night are 12 hours long. This occurs because the Sun is then directly on the celestial equator. The vernal (spring) equinox is on about 21 March each year; the autumnal equinox is on about 22 September.

F

Front: the boundary between two air masses which differ in temperature or density. On a warm front, warm air replaces cold; on a cold front, cold air replaces warm.

G

Galaxy: a star system. There are two types: 1 spiral such as our own Galaxy (the Milky Way) and the Great Andromeda Nebula (M31), which contain dust and gas; and 2 elliptical galaxies, which are composed almost exclusively of stars.

Gas giant: one of the four planets – Jupiter, Saturn, Uranus and Neptune – with a high percentage of gaseous or liquid components around a comparatively small solid core.

Gravity: a force between all bodies that attracts one to the other. It is quite distinct from magnetic attraction.

H

Halo: an optical phenomenon caused by refraction (or reflection) of light from the Sun or Moon. The most common effects are rings with radii of 22° and 46°. (NB: it is not an indicator of bad weather.)

Heiligenschein (holy light): the white halo sometimes visible around the shadow of an observer's head, usually when it falls on dew-covered grass. It is caused by the diffraction of light by water drops.

Hertzsprung-Russell diagram (H-R): a chart showing the relationship between the luminosity of stars and their spectral type or colour.

I

Inferior planet: a planet whose orbit lies between that of the Earth and the Sun, eg Mercury and Venus (see *Superior planet*).

Ionosphere: the layer of the Earth's atmosphere between 56 to 89 km (35 to 55 mls) and about 805 km (500 mls) above sea level. Free, electrically charged particles in this region enable radio waves to be transmitted great distances around the Earth. The structure of the ionosphere alters constantly as a result of the Sun's influence.

J

Jet stream: a narrow belt of strong winds, only a few kilometres deep but usually hundreds of kilometres wide, near the troposphere. Minimum wind speeds exceed 100 km/h (62 mls/h).

L

Latent heat: the heat required to convert a solid into liquid or vapour, or a liquid into vapour, without change of temperature.

Light-year: a unit of measurement equal to the distance travelled by light in one year when moving in a vacuum. This is 9.4607×10^{12} km or 63,240 *astronomical units*.

Limb: the edge of the visible disc of a body such as the Sun.

M

Magnitude: a measurement of brightness calculated in a variety of ways:
1 Apparent magnitude describes the degree of brightness of an object as seen from the Earth. Very bright stars are considered to have a negative magnitude (e.g. Sirius − 1.46).
2 Absolute magnitude overcomes some of the distortions caused by the distance of objects in estimating their relative brightness. It records the apparent magnitude they would possess if lying at a distance of 10 *parsecs*.
3 Bolometric magnitude measures the total amount of radiation emitted by a star, not solely the fraction that constitutes visible light.

N

Nova: a star, usually a member of a binary system, which suddenly flares up, emitting a layer of hot gas, and so briefly appears extremely bright. A supernova is a much more violent explosion of a star at the red supergiant stage or of a white dwarf star in a binary system.

O

Occultation: describes the moment at which one celestial body passes in front of another, temporarily obscuring it.
Opposition: when a superior planet is on the opposite side of the Earth to the Sun, it is said to be in opposition. This is the best time to observe the planet, for at midnight

local time it will be due south.
Ozone: a form of oxygen containing three atoms in each molecule. It is found in minute amounts near the Earth's surface and in larger quantities 20 to 50 km ($12\frac{1}{2}$ to 31 mls) above the Earth's surface.

P

Parallax: the apparent shift of an object across the sky created by the actual movement of the observer. The distance between the two points from which the object is seen is called the baseline. The diameter of the Earth's orbit provides a baseline of almost 300 million km (186 million mls). This is used as a method of measuring the distance of a star because the nearest stars show a larger parallactic shift than stars farther away. These shifts are only detectable to a distance of 300 *light-years*.
Parsec: a unit of measurement equal to the distance at which a star would be positioned to have a parallax of one *arc second*. The Sun is the only star to lie within one parsec (pc) of the Earth. A parsec is equal to 3.26 *light-years*, 206,265 *astronomical units*, or 30.857×10^{12} km (19.174×10^{12} mls).
Penumbra (Umbra):
1 The area of partial shadow, either of the Moon (in a solar eclipse) or of the Earth (in a lunar eclipse), distinct from the Umbra, the area of total shadow. **2** The pale outer area of a sunspot.

Perihelion: the closet point of an orbiting body to the Sun (see *Aphelion*).
Photosphere: the luminous apparent surface of the Sun or other star.
Planetary nebula: a shell of bright gas surrounding a star.
Planisphere: a flat, circular star map overlaid with a circular window that can be turned to show the stars visible at any time on every night of the year. They are available for both southern and northern skies.
Pore: a minute dark spot on the Sun; the smallest *sunspots*.
Prism: a glass figure (often triangular) with the sides inclined to one another. Light passing through a prism is split into its component rainbow colours. Erecting prisms are used in binoculars to invert an upside-down image.
Prominence: a mass of gas, arising from the chromosphere of the Sun, that appears as a bright cloud or streamer.

Q

Quasar: a quasi-stellar object (QSO); an intensely luminous celestial body, often emitting considerable radio radiation. Quasars are probably extremely distant and it is likely they are the central regions of highly active galaxies in an early stage of development.

R

Radiotelescope: an instrument for studying the radiation at radio wavelengths emitted by celestial objects.

Refraction: the deflection of light rays from a straight path when they pass obliquely from one transport medium (eg, air) through another (eg, water or glass) in which their velocity is different.

Regolith: a layer of broken rock and rock dust on the Moon and Mars, caused by the bombardment by meteors.

Relative humidity: the amount of water vapour in the air expressed as a percentage of the amount it could contain at the same temperature if fully saturated.

Right Ascension (RA): the equivalent on the celestial sphere of terrestrial longitude. It is measured in hours, minutes and seconds (1 hour = 15°) eastward along the celestial equator. The prime meridian runs North-South through the first point of Aries, and the difference in time between this point and the position along the celestial equator of a celestial body gives the right ascension of that body (see *Declination*).

S

Sidereal period: the orbital period of a planet, satellite, asteroid or comet measured from one star to its return to that star.

Sidereal time: a method of measuring time based on the rotation of the Earth in relation to the stars, rather than to the Sun (solar time). The sidereal day is approximately four minutes shorter than the solar day and begins when the vernal equinox crosses the observer's South point or meridian.

Spectroscope: an instrument used to observe and analyze the light and other radiation emitted by luminous objects.

Stratosphere: the layer of the atmosphere above the *troposphere* that extends from about 11 to 64 km (7 to 40 mls) above the Earth.

Sublimation: the process by which a solid changes to a gas without passing through a liquid state.

Subsun: a *halo* phenomenon in the form of a bright spot of light at the *antisolar point* produced by reflection from ice crystals.

Sunspots: relatively cool areas of the Sun's surface, which appear dark by contrast. They tend to occur in the equatorial regions and may last for a few weeks or, exceptionally, a few months.

Superior planet: a planet whose orbit lies farther from the Sun than the orbit of the Earth – eg, Mars, Jupiter, Saturn, Uranus, Neptune and Pluto (see *Inferior planet*).

Synodic period: the length of time between one particular alignment of a celestial body and the next occurrence of that alignment.

T

Terminator: the division across a planet or satellite separating the sunlit and dark hemispheres.

Transient Lunar Phenomena (TLP): observable features over the surface of the Moon which are constantly changing and may obscure some features with colorations or brightenings.

Transit: a planetary configuration where a celestial body either crosses the disc of another, eg, the transit of Mercury across the Sun, or crosses a specific point, eg, the transit of a star across the Equator.

Terrestrial planet: one of the four relatively small and solid planets orbiting closest to the Sun. These are Mercury, Venus, Earth and Mars.

Troposphere: the lowest part of the Earth's *atmosphere*, ranging from 8 to 18 km (5 to 11 mls) in depth, in which temperature generally declines with altitude and where the majority of clouds is found.

V

Variable star: a star whose brightness constantly changes. This may be due to pulsation – the expansion and contraction of its surface. Some variables are binary systems in which two non-pulsating stars periodically eclipse one another.

W

White dwarf: a star of comparable size to the Sun but close to the end of its life; it shines only from its surface which has become white-hot. Its central regions are extremely dense.

Index

A

Absolute magnitude 140–1, 215
Achenar 159
Adams, John Couch 112
Aerial perspective 17
Air
circulation 30–4
cooling 30, 42
humidity 40, 42
masses 36
mountain waves 54
orographic lifting 43
pressure 10, 12–13, 30, 32, 34
Aldebaran 136, 137, 149, 158, 163
Alexander's dark band 23
Algol 168
Altair 159
Altazimuth mount 198
Altocumulus (Ac) 44, 52–4, 59, 62, 64
Altostratus (As) 44, 50, 60
Andromeda (M31) 126–7, 130, 147, 149, 166
Anemometer 35, 36, 94–5
Aneroid barometer 12
Ankaa 147
Antares 155
Anticyclones 34
Antoniadi scale 202
Anvil clouds 56
Apparent magnitude 140–1, 215
Aquarius 146, 156, 158, 162
Aquila 133, 157
Arctic (A) air masses 36
Arcturus 158,
Aries 147, 158
Asteroids 103
Astrology 132–3
Astronomical unit (a.u.) 105, 208
Astrophysics 208
Atlases, star 142–3
Atmosphere
density 14
extent 10–11
gases in 12
temperature variations 10–11
water vapour in 12
Atmospheric haze 18, 186
Atmospheric particles 16, 18
Atmospheric pressure 10, 12–13, 32
Aureole 24, 26
Auriga (M37/8) 133, 148, 158, 214
Aurorae 74, 82–3
Axial rotation 100–3

B

Baily's Beads 76
Balloons, meteorological 38–9
Banner clouds 42–3
Barograph 13, 92
Barometers 10, 12–13, 92
Barred spirals 130
Beaufort Scale 35, 36, 94
Beehive (M44) 164, 214
Betelgeuse 120, 123, 135, 136–7, 141, 158
Big Bang theory 131
Binary stars 166–9
Binoculars
choosing 194–5
limiting magnitudes 209
magnification 192–3
observing with 46, 163–4, 179, 214
types 192
Black holes 121, 123
Bolometric magnitude 141
Boötes 154, 158
Brightness ratio 208
Brocken Spectre 26, 28
Butterfly diagram 73
Buys-Ballot's law 35

C

Callisto 108, 211
Cancer (M44) 151, 214
Canes Venatici (M3) 152, 164, 166, 214
Canis Major 137, 151, 158
Capella 148, 158
Capricornus 133, 166
Carina 137, 158
Cassegrain reflecting telescope 196, 197
Cassini division 111
Cassiopeia (M103) 136, 158, 215
Castor 120, 138, 166
Celestial equator 134
Celestial latitude 142
Celestial longitude 142
Cellarius, Andreas 133
Centaurus (NGC 3766) 137, 155, 158, 214
Cepheid variables 169
Cepheus 136, 143, 158, 166, 215
Ceres 103
Cetus 159
Charon 113, 211
Chromosphere 76
Circulation patterns 30
Circumpolar constellations 143–4
Circumzenithal arc 28
Cirrocumulus (Cc) 44, 53, 54, 62

Cirrostratus (Ce) 28, 44, 52, 59, 60, 80
Cirrus (Ci) 28, 31, 56, 58, 59, 60, 62, 80, 90, 186
Clouds
airborne observation 84–9
bands 54
banner 42–3
colour 18, 26, 50, 52
coronae 24
glory 26, 28
height 44
iridescent 26–7, 80
lenticular 54–5
nacreous 80
noctilucent 80–1
photographing 186
pileus 51
rain-bearing 50, 56–8
stratospheric 80
symbols 44–5
and thunderstorms 64
tropospheric 12, 80
wave 54
as weather indicators 50, 60, 90
See also individual cloud types
Clusters 160–5
Comets 103, 161, 174–5, 190
Condensation, 86, 88
Constellations
coordinates 142
designation 132–3
English/Latin names 212–13
maps 140–59
Contrails 60, 86, 88–9
Convection 30, 40–1, 84
Coordinates 134, 142
Cor Caroli 152
Coriolis effect 30
Corona (auroral) 24, 26, 82
Corona (solar) 74, 76–7, 82
Corona Australis 157
Corona Borealis 154
Corvus 153, 159
Crater 152, 159
Crepuscular rays 17
Crux (NGC 4755, Southern Cross) 137, 158, 159, 164, 214
Cumulonimbus (Cb) 47, 48, 56, 58, 60, 62, 64, 66, 84
Cumulus (Cu) 40–1, 44, 46, 84, 186
Cyclones 34, 60
Cygnus (M39) 133, 157, 158, 166, 214

D

Declination 134, 142, 215
Deimos 107, 211

Delphinus 167
Depressions 34, 37, 38, 60–9
Diamond Ring effect 76
Diffraction 24
Digges, Thomas 124
Dione 111, 211
Dispersion 14
Double Cluster 164
Double stars 166–9

E

Earth
  polar axis 103
  rotation 30, 32, 100–1
Eclipses
  lunar 114–15, 182, 210
  solar 74–7, 114–15, 210
Eclipsing binaries 168
Ecliptic 142
Elliptical galaxies 129–30
Equator 30–2
Equatorial drive 190, 198–9
Equinox 134, 142
Europa 108, 211

F

Faculae 72
Fallstreaks 53, 59
Fata Morgana 78
Filaments 74
Filters, photographic 184–5
Fog 34, 49–50, 86
Fogbow 22
Fomalhaut 146, 162
Fronts 36–8, 60–3, 84–6

G

Galactic latitude 142
Galactic longitude 142
Galaxies
  barred spiral 130
  clustering 130–1
  elliptical 129–30
  quasars 131
  spiral 129–30
  See also individual
    galaxies
Galileo 125
Ganymede 108, 211
Gemini (M35) 133, 136, 138,
  151, 166–7, 168, 214
Globular clusters 162–3,
  164–5
Glory 26, 28, 86
Gravitational pull 121
Great Comet 173
Great Red Spot 109
Greek alphabet 120
Greenwich Mean Time
  (GMT) 96, 203

H

Hail 66
Halley, Edmond 124

Halley's Comet 175
Halo 24, 28, 52, 60, 86, 88,
  90
Heiligenschein 86
Helix nebula 162
Hercules (M13, Great
  Cluster) 143, 154, 165,
  214, 215
Herschel, William 112, 125,
  162, 166
Hertzsprung, Enjar 121
Hertzsprung-Russell
  diagram 122
Hipparchos 138–9
Hoar frost 39
Horsehead nebula 160
Humidity 40, 42
Hurricanes 68
Hyades 149, 163
Hygrometer 40, 92–3

I

Iapetus 111, 211
Ice crystals 28–9, 56–8
Index Catalogues (IC) 161
Inferior mirage 78 9
International
  Astronomical Union
  (IAU) 134
Io 108, 211
Isobars 32

J

Jet streams 34
Jupiter
  atmospheric currents
    108–9
  brightness 108, 110
  future positions, 205, 207
  Great Red Spot 109
  mass 108
  observation 108–9, 159,
    202
  polar axis 103
  rotation 109
  satellites 108, 211

K

Katabatic winds 34
Kohoutek 175
Kruger 60, 166

L

Lagoon 161
Lenticular clouds 54–5
Leo 133, 140, 151, 153, 158,
  170
Leo Minor 151, 153
Leonid meteorites 170, 173
Lepus 148, 150, 159
Leverrier, Urbain 112
Libra 154
Light 14, 16, 18, 24, 26–8,
  86, 208

Light-years 126, 208
Lightning 64–6, 90, 186
Local Group 130
Low-pressure zones 30, 32,
  84
Lunar eclipses 114–15, 182,
  210
Lupus 133, 145, 155
Lyra 133, 157, 158, 162, 166,
  168, 215

M

Magellanic Clouds 131,
  136, 159
Mare's tail 56, 90
Maria 118
Maritime (m) air masses 36
Mars
  future positions 204, 206
  observation 106–7, 159,
    202
  orbit 106–7
  satellites 107, 211
Mercury
  future positions 204, 206
  observation 104–5, 202
  orbit 104–5
  transit 105
Messier, Charles 161
Meteorites 170–5, 190, 213
Meteorological balloons 39
Meteorological stations
  92–7
Microscopium 133
Milky Way 124–8
Millibars 10
Mira 148, 169
Mirages 78–9
Mistral 36
Mitchell, John 166
Mizar 166
Mock suns 28
Monoceros 162
Moon
  aureole 24, 26
  axial rotation 119
  craters 118
  diameter 114
  eclipse 114–15, 182, 210
  gravitational field 114
  halo 28
  map 116–17
  marias 118
  observation 115–16,
    118–19
  orbit 114
  rising 182
  surface details 115,
    118–19
  synodic months 114
  Transient Lunar
    Phenomena 118–19
Moonlight 22, 120

N

Nacreous clouds 80
Nebulae 160–5
Neptune
  axis of rotation 112
  discovery 112
  future positions 205, 207
  observation 113
  satellites 211
Neutron stars 123
New General Catalogue
  (NGC) 161
Nimbostratus (Ns) 44, 50,
  58, 60
Noctilucent clouds 80–1
Nova 167

O

Observation
  from air 84–9
  with naked eye 182–3,
    214, 215
  records, keeping 202–3
Observatory, home 200–1
Open clusters 162–3
  Orbits
  comets 175
  planets 102
  satellites 177–9
  stars 124–5
Orion 136–7, 140, 141, 158,
  159, 160, 167, 215
Orographic lifting 43
Ozone layer 12

P

Parallax 126, 128
Parhelia 28
Parsec 126, 208
Particle size 16
Pegasus 165, 214
Perihelion 175
Perseus 148, 158, 160, 168,
  214, 215
Phobos 107, 211
Phoenix 159, 168
Photography
  daylight 184–7
  night 188–91
Photosphere 70, 76
Pileus 51
Pisces 147
Planetary nebulae 162
Planets
  axial rotation 102–3
  diameters 100
  distances from Sun 100
  future positions 204–7
  orbits 101, 102
  rotation 100–1
  satellites of 211
  See also names of
    individual planets

Planisphere 135, 158–9
Pleiades (Seven Sisters)
  136, 149, 162–3, 214
Plough (Dipper) 143, 158,
  166
Pluto
  discovery 112
  future positions 205, 207
  observation 113
  orbit 113
  satellite 211
Polaris (Pole Star) 134, 136,
  143
Poles
  celestial 134, 159
  and air flow 30, 32
Pollux 120, 138, 164
Population I stars 130, 162
Population II stars 130
Porro prism 192
Praesepe 163, 214
Prominences 70–2, 74
Puppis 137, 151

Q

Quasars 131

R

Radiosonde 38
Rain
  clouds 48, 50, 56–8
  droplets 22
  formation 57–8
  gauge 93–4
  -making experiments 42
  measurement 92–4
  symbols 57
Rainbows
  Alexander's dark band 23
  colours 20, 22
  formation of 20–2
  primary bow 20, 23
  reflected bow 22
  secondary bow 20
  seen from air 86
  supernumerary bow 20,
    22
  as weather indicator 22,
    24
Records, keeping 92–7,
  202–3
Red giants 123
Refraction 14, 20–9
Regulus 153, 158, 164
Relative humidity 40, 42
Rhea 111, 211
Rigel 120, 135, 136–7, 141,
  158
Right ascension (RA)
  134–5, 142, 215
Rigil Kentaurus 145, 159
Roof prism 192
Rosette nebula 162

Rotation, planetary 100–3
Russell, Henry Norris
  121–2

S

Sagittarius 128–9, 133, 157,
  161, 162, 165, 214
Satellites
  astronomical 176–7
  communications 177
  IRAS 160
  observation 177–8
  photographing 189–90
  planetary 107, 108, 111,
    211
  research 176–9
  weather 38, 176
  X-ray detecting 167, 176
Saturn
  atmospheric currents 11(
  brightness 110
  density 110
  future positions 205, 207
  observation 110–11, 159,
    202
  rings 110–11
  satellites 111, 211
Scattering 14, 16, 18
Scintillation 14–16
Scorpio 133
Scorpius 137, 155, 214
Scutum (M11) 162, 214
Serpens Caput 154–5, 165
Shooting stars 170–5
Sirius 108, 140, 158
Sleet 58
Snow 58
Solar System 102–3
Speckle interferometer 12:
Spectral shift 131
Spectroscope 125
Spectrum 15, 20, 120–1, 131
  167
Squall line 66
Square of Pegasus 126, 136
  146, 158
Stars
  binary 166–9
  catalogues 161
  catalogue numbers
    139–40
  charts 135, 136, 142–59
  classification 122, 132–3,
    139–41
  colour 120, 167
  designations 120
  distances, measuring 12(
    128
  diurnal motion 100
  double 166–9
  'fixed' 124
  gravitational pull 121
  life cycle 122–3, 130

magnitudes 138–41
movement 124–5
observation 158–9
parallax 126, 128
positions 142
scintillation 14–16
shooting stars 170–5
spectral type 122
spectroscope 125
variable 169
Stop-watch 178
Stratocumulus (Sc) 44,
  50–2, 62
Stratosphere 12
Stratus (St) 44, 48–50, 62
Sun
  aureole 24, 26
  auroral activity 82
  chromosphere 76
  corona 74, 76–7
  Diamond Ring effect 76
  distance of planets from
    100, 126, 128
  eclipse 74–7, 114–15, 210
  energy 70
  faculae 72
  filaments 74
  galactic orbit 124–5
  granulation 72
  halo 28
  limb 70
  magnetic field 74
  observation 24, 70, 76,
    184, 202
  penumbra 70, 72
  photosphere 70, 76
  pillar 28
  planetary orbits 102
  prominences 70–2, 74
  size 70
  solar cycle 73
  solar wind 70–7, 82
  sunrise 16, 20
  sunset 16, 18, 20
  sunshine recorder 95
  sunspots 70–3
  temperature 70, 72
  umbra 70, 72
Supergiants 123
Superior mirage 78
Supernova 167
Synodic month 114
Syrtis Major 106

T

Taurus (M45) 133, 136, 149,
  158, 163, 214
Telescopes
  caring for 197
  drives 190, 198
  eyepieces 196
  excessive magnification
    163

limiting magnitude 209
mounts 198–9
observing with 161, 162,
  163, 164, 165, 166–9, 179
reflecting 196–7
refracting 196, 200
viewfinders 198
Telescopium 133
Tethys 111, 211
Thermals 46, 84
Thermohygrography 95
Thermometers 42, 92–4
Thunderstorms 64, 66
Tirion, Will 136
Titan 111, 211
Tombaugh, Clyde 112
Tornadoes 66–8
Towering 78
Transient Lunar
  Phenomena 118–19
Triangulum 158
Trifid nebula 164
Tropical (T) air masses 36
Tropopause 12, 44, 80
Troposphere 10–13, 44, 80
Troughs 32, 34
Tucana 159, 167, 214
Turbulence 84
Tycho 118

U

Ultraviolet radiation 12
Universal Time (UT) 96,
  203
Uranus
  axis of rotation 112
  discovery 112
  future positions 205, 207
  observation 112
  rings 113
  satellites 113, 211
Ursa Major 136, 158, 167
Ursa Minor 136, 166, 167

V

Variable stars 169
Vega 157, 158, 159, 162
Vela 137, 145, 151, 158
Venus
  distance from Sun 105
  future positions 204, 206
  observation 104–5
  orbit 104–5
  polar axis 103
  transit 105
Virga trails 53, 59
Virgo 131, 153
Volans 145, 159
Vortices 84, 88

W

Wave clouds 54
Wave cyclone 34, 37, 38,
  60–9

Water
  in atmosphere 40–3
  spouts 68
  vapour 12, 40
Weather
  forecasting 10, 22–4,
    38–9, 90–1
  fronts 36–8, 60–3, 84–6
  isobaric charts 32
  lore 90–1
  map 37
  observation 92–7
  records 92–7
  stations 92–7
Whirling hygrometer 92–3
Whirlwinds 68
White dwarfs 123
Wind
  and air/land temperature
    34
  Beaufort Scale 35, 36, 94
  direction 30–1
  jet streams 34
  measurement 35–6, 94
  and monsoons 34
  movements 30–9
  and pressure differences
    32
  shear 54
  tornadoes 66–8
  in valleys 34
  whirlwinds 68
World Meteorological
  Organization (WMO)
  38

Z

Zodiac 132–3, 208

Acknowledgements

Artwork illustrations: David Parker; Hayward and Martin

Index: Tim Shackleton

Many people have helped in the production of this book, and while it is not possible to thank them all individually, we should especially like to mention Paul Doherty, Richard McKim, Tony Sizar, Peter Hingley, Librarian of the Royal Astronomical Society, and the staff at the British Astronomical Association. We are also grateful to the British Astronomical Association for permission to reproduce part of a Will Tirion Star Chart; to Bill Fox and Alan Heath for the provision of observers' record cards, and to Kodak Limited for photographic information.

Picture credits
Key: **b** bottom **t** top
Endpapers Chuck Wise/AA/OSF; **1** Paul Dix/Susan Griggs Agency; **2/3** Sherman Hines/Daily Telegraph Colour Library; **4/5** Hans Blohm/Daily Telegraph Colour Library; **6/7** Tom Grill/Photofile Int.; **8/9** John Garrett; **11** Daily Telegraph Colour Library; **13** Casella London Ltd; **14** P. Freytag/Zefa Picture Library; **15** David Parker/Science Photo Library; **17t** Munzig/Susan Griggs Agency; **17b** John Sims; **19t** John Cleare; **19b** Barrie Rokeach/The Image Bank; **20/1** Armstrong/Zefa Picture Library; **22/3** Douglas W. Johnson/Science Photo Library; **25** James Davis; **26/7** Doug Allan/Science Photo Library; **27** Helmut Gritschee/Aspect Picture Library; **29** Doug Allan/Science Photo Library; **30/1** R. K. Pilsbury; **35** Casella London Ltd; **37** Crown Copyright/Meteorological Office; **39t** Simon McBride/Susan Griggs Agency; **39b** John Yates/Topham Picture Library; **41t** Tony Stone Assoc.; **41b** Storm Dunlop; **42/3** Reflejo/Susan Griggs Agency; **46/7** Sean Morris/Oxford Scientific Films; **49** Dr E. I. Robson/Science Photo Library; **50/1** Walter Rawlings/Robert Harding Picture Library; **51** Ingrid Holford; **52/3** Storm Dunlop; **53** R. K. Pilsbury; **54/5** Dr R. Spicer/Science Photo Library; **56/7** Maurice Nimmo/Frank Lane Picture Agency; **58/9** C. Carvalho/Frank Lane Picture Agency; **63** NASA/Aldus Archive; **65** T. Ives/Zefa Picture Library; **66** NCAR/Science Photo Library; **67** Robert McNaught; **68** Jim Howard/Colorific!; **69** Tony Stone Assoc.; **70** Lockhead Solar Observatory/Aldus Archive; **71** Daily Telegraph Colour Library; **72** P. Parviainen/Finland; **74/5** Dr Fred Espenak/Science Photo Library; **76/7** John Fairweather; **78/9** Pierre Boulat/Colorific!; **80/1** R. K. Pilsbury; **82/3** Jack Finch/Science Photo Library; **85** John Cleare; **86/7** Mike McQueen/The Image Bank; **87** Morton Beebe/The Image Bank; **88/9** Walter Rawlings/Robert Harding Picture Library; **89** Robert McNaught; **90/1** M. Garff/The Image Bank; **98/9** John Cleare; **101** Daily Telegraph Colour Library; **104/5** Lowell Observatory Photograph/Aldus Archive; **105** Jet Propulsion Laboratory/Aldus Archive; **106/7** NASA/Science Photo Library; **109t** Paul Doherty; **109b** Daily Telegraph Colour Library; **110** NASA/Science Photo Library; **112t** Paul Doherty; **112b** Lowell Observatory Photograph/Aldus Archive; **113** Lowell Observatory Photograph/Aldus Archive; **114/15** Lick Observatory Photographs; **115** Daily Telegraph Colour Library; **116** NASA/Aldus Archive; **116/17** Lick Observatory Photograph; **119** NASA/Aldus Archive; **123** National Optical Astronomy Observatories/Kitt Peak; **124/5** Lund Observatory; **127** California Institute of Technology & Carnegie Institution of Washington/Aldus Archive; **128/9** Royal Astronomical Society; **130/1** Royal Observatory, Edinburgh; **133** The British Library; **134/5** N. R. Patel, A. J. Sizer/Chigwell School Astronomical Society; **135** Hatfield Polytechnic Observatory; **138/40** N. R. Patel, A. J. Sizer/Chigwell School Astronomical Society; **141** Hatfield Polytechnic Observatory; **160** Lick Observatory Photograph; **161** Royal Observatory, Edinburgh; **162** Daily Telegraph Colour Library; **163** Royal Astronomical Society; **164/5** Royal Astronomical Society/Aldus Archive; **171** Pete Turner/The Image Bank; **172** Michael Holford; **173** Ann Ronan Picture Library; **174/5** Lowell Observatory Photograph/Royal Astronomical Society; **179** NASA/Daily Telegraph Colour Library; **180/1** Tony Stone Assoc.; **182** Martin Dohrn/Science Photo Library; **183** Michael Freeman/Bruce Coleman; **184/5** Pisces Pics; **186/7** NCAR/Science Photo Library; **187** John Garrett; **188** Adam Woolfitt/Susan Griggs Agency; **189** Michael J. Hendrie; **190/1** Harvard College Observatory/Science Photo Library; **191** R. Royer/Science Photo Library; **194/6** Lick Observatory Photographs